BUYING, SUPPORTING, MAINTAINING SOFTWARE AND EQUIPMENT

An IT Manager's Guide to Controlling the Product Lifecycle

BUYING, SUPPORTING, MAINTAINING SOFTWARE AND EQUIPMENT

An IT Manager's Guide to Controlling the Product Lifecycle

Gay Gordon-Byrne

CRC Press
Taylor & Francis Group
Boca Raton London New York

CRC Press is an imprint of the
Taylor & Francis Group, an **informa** business

AN AUERBACH BOOK

CRC Press
Taylor & Francis Group
6000 Broken Sound Parkway NW, Suite 300
Boca Raton, FL 33487-2742

First issued in paperback 2019

© 2014 by Taylor & Francis Group, LLC
CRC Press is an imprint of Taylor & Francis Group, an Informa business

No claim to original U.S. Government works

ISBN-13: 978-1-4822-3278-3 (hbk)
ISBN-13: 978-1-4822-3278-3 (hbk)

Library of Congress Cataloging-in-Publication Data

Gordon-Byrne, Gay.
   Buying, supporting, maintaining software and equipment : an IT manager's guide to controlling the product lifecycle / author, Gay Gordon-Byrne.
      pages cm
   Includes bibliographical references and index.
      ISBN 978-1-4822-3278-3 (hardback)
   1. Computer systems--Management. 2. Computer systems--Purchasing. 3. Software maintenance. 4. Product life cycle. I. Title.

QA76.5.G633 2014
005.1'6--dc23
                                                                        2014007194

Visit the Taylor & Francis Web site at
http://www.taylorandfrancis.com

and the CRC Press Web site at
http://www.crcpress.com

Contents

SECTION II   Postwarranty
Support and Maintenance

SECTION III Technology Product Details

SECTION IV   Controlling Product Life

Preface

Welcome. This book is not a how-to manual for repair nor is it a how-to negotiate prices manual. My goal is to help end users understand why vendors propose the service agreements they offer. Readers will be able to avoid common traps, better control the negotiation for both products and services, and thereby control the lifecycle of their purchases.

This book is set up to follow the way that most people experience technology products and contracting decisions. The first section deals with decisions made at the time of product selection, then examines the types of problems typically experienced during use, and finally delves into issues of product end of life, and how best to manage support and maintenance issues for the long term.

Readers may notice that many of my concepts and recommendations for negotiation will not be welcomed by manufacturers. This is not accidental. My lengthy career as a vendor has given me wide experience with the methods and goals of vendor contracts, which are almost always in conflict with end-user goals. I am standing on the side of end users, fighting against my former self. Enjoy.

Section I

Initial Product Acquisition

1

Equipment and Application Acquisition

INTRODUCTION

This chapter focuses on the relationships between choices of software applications, operating systems, and hardware from the standpoint of the manufacturer goals in the transaction(s). Buyers are often unaware of the monopolistic lock-ins and account control intentions of the providers, with the result that many purchases are not negotiated to the advantage of the buyer.

APPLICATION SELECTION DICTATES HARDWARE SELECTION

Acquisition of technology products follows the application. At the consumer level, individuals buy products for their function, not the design of the circuit board. Businesses, industry, and government select the application software first, and then the hardware that goes with it. Information technology (IT) managers are often pressed to look at "technology" as a solution to business problems; however, the real fact is that the solution to business problems is an application, not technology. It does not matter how the bits and bytes are configured or if the processor capability is less than leading edge; if the application does not support the business.

It is only after a commitment has been made to a particular application that the hardware platform and associated operating system become meaningful. The long-term useful life of the equipment is tied to the planned useful life of the application. If the application decision is in error, the associated hardware and service contracts will also gather dust.

This chapter is directed at unveiling nuances of lifecycle control points created by vendors that can only be effectively mitigated before the initial purchase. Careful attention to the initial product selection that goes beyond the application negotiation and into the tangled support functions of hardware, application license, and operating system (OS) license will bring long-term rewards.

The selection of products and associated support is a constant tug-of-war between the buyer and the manufacturer, usually known as the original equipment manufacturer, or OEM. The buyer and the seller are not involved in a traditional purchase agreement, despite the appearance. Buyers do not own IT equipment in the same manner as they own a house or even as they own a vehicle. Nor is the license agreement for software a straightforward arrangement like a condo lease. The relationship between buyer and seller, or licensee and developer, is more like a marriage than a home purchase. Buyers should be negotiating IT purchases and license agreements just as carefully as they would a prenuptial agreement expecting divorce and conflict.

Figure 1.1 shows the OEM view of the sale or license agreement.[1] OEMs expect the user to enter into a long-term relationship where the OEM controls the useful life of the product with no exit points for the user. Although OEMs might not set out to produce products that are unstable and require repairs and software patches, they have little incentive to do otherwise. It is beneficial for OEMs to have reasons for the user to need their support.

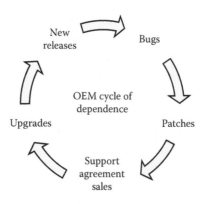

FIGURE 1.1
OEM cycle of dependence. (Adapted from www.continuant.com. With permission.)

Background

Manufacturers of hardware long ago recognized that the application drove the hardware. Outside of the mainframe environment, where initially most applications were custom built, the entire middle range of product sales were built around what used to be called "Turnkey" systems. The turnkey arrangement was perfect for the business that had no computerized applications—and thus no employees—capable of writing, customizing, or implementing the application. Most of the big names in applications today came from these roots.

Hardware manufacturers were not initially in the application software business. Nor did hardware companies sell or separately license the operating system. In order to sell to the neophyte buyer, hardware manufacturers made alliances with software developers, later called business partners or value-added resellers, to represent them to new or smaller users. Software vendors became the sales force for their own products and a free extension of the OEM sales force as an authorized partner. The software company got a healthy commission for selling the hardware. The operating system and the requisite maintenance contracts were also sold, and commissions paid, to the application partner.

It is for these historical reasons that the sales channel through which equipment is purchased usually dictates how both hardware maintenance (break–fix) and software maintenance (operating system and application support) are offered. This chapter discusses how different players in the supply chain have different offerings and how negotiations should be conducted to maximize effectiveness in support options.

Each type of acquisition method still involves navigating the natural tension between seller and buyer. The seller always wants to sell more equipment, more quickly, and drag along higher margin services at the highest possible price point. The buyer has goals of buying equipment that best supports the application requirements, at the best price, and with the least likelihood to need replacement ahead of the depreciation schedule.

Application License Acquisition

Buyers are commonly exasperated with trying to control maintenance costs for software, the price of which often exceeds the acquisition cost of the license over time. The reason is very simple: vendors have total control of the pricing of postwarranty support (maintenance). As with any other monopoly, this ability to dictate price and terms leads inevitably to taking advantage of the profit margin potential of the revenue stream. Maintenance is often the most profitable part of the license agreement. It is therefore essential that the initial negotiation for application licenses be extremely attentive to issues that might occur only in the far future.

There are only a handful of ways to control maintenance pricing, at the outset, through competitive replacement or a commitment to open systems. None of these tactics are perfect. Applications are completely protected by copyright law (which is itself consistent with both the Berne Convention and World Intellectual Property Organization [WIPO] international treaties). As "Intellectual Property (IP)," vendors are allowed to be monopolistic not only about the transfer of licenses but also about ongoing support. There is no legal (as of this writing) opportunity for competition for support unless the application vendor allows it.

Unfair terms and conditions in the "End User License Agreement (EULA)" are coming under scrutiny in general as digital rights are being considered. Copyright law, including international conventions, remains rooted in the predigital era. Even legislation passed at the end of the millennium to update the Copyright Code under the DMCA (Digital Millennium Copyright Act of 1998) seems archaic and dated. Some work has been done defining conceptual "User Rights" by the Gartner Group in 2010, none of which has been put into law.[2]

The ideal time to control maintenance costs in the future is to carefully negotiate the initial contract to include limits on future increases for maintenance costs. These negotiations usually focus on a percentage of the original list price, or original acquisition cost, rather than the need for maintenance in the future. The downside of this approach is that buyers are stuck with whatever percentages they negotiate, on top of which application vendors can inflict further financial burdens for upgrades to major "new" releases, requirements that all versions of the purchases be kept current, or other requirements such as linkages to hardware maintenance agreements.

The concept of the responsibility of the developer to deliver bug-free code is lost in the focus on discounting. Vendors have been allowed to own the negotiation over support and limit the discussion to discounting off list price for future support contracts. This is exclusively to the advantage of the vendor and always puts the buyer in the position of begging for discounts. The framework of the negotiation can change if buyers demand performance-based postwarranty support in the initial license negotiation. Once the focus of the negotiation is put on quality and performance on the part of the vendor, the entire spectrum of costs and rights can be discussed.

Defect and performance-based support is rooted in the idea that the vendor, either hardware or software, has an obligation to deliver fully functional products that meet the specifications as advertised. Patches and fixes are indications that code is buggy. Users need to stop acquiescing to the concept that buyers should pay for corrections to code that should not need correction.

Users have been led to believe that they should pay handsomely for support (maintenance) of flawed code because the patches and fixes are delivering improvements in the product. This is unlikely to be the case, although there may be exceptions. There is more than a semantic difference between an "update" and an "upgrade." An *update* is likely to be double-speak for a bug fix, but sounds far more positive. An upgrade should provide a new and valuable function that did not previously exist. An upgrade should be something that the user might want to purchase, separately. Sadly, many software and hardware vendors blithely mix the wording of updates and upgrades simply to hide the weaknesses in their product stability.

The costs and rights for users to buy patches and fixes to flawed code can, and must, be separated from enhancements and customizations. Users can begin the process of accountability by evaluating the patches and fixes that are disseminated for existing licenses for applicability, impact, and separating upgrades from patches. Much of this work is already being done by support staff to discern which patches must be urgently applied. Adding the layer of intelligence to differentiate between welcomed and unwelcomed updates will enormously increase the negotiating power of the user in future contracts.

Going Naked

It is an underutilized but viable option to drop maintenance (support) where no meaningful changes are needed and freeze the system, both hardware and software, at a stable release. Vendors use the term "going naked" to infer the undesirable vulnerability of dropping vendor support. The key to confidence in dropping "support" is to evaluate the types of patches and fixes being issued and consider if the improvements being made are valuable. Only the user can judge if the problems being fixed are meaningful.

In many cases, application systems patching follows the pattern of hardware systems patching. The initial release is full of bugs, which require a high level of interaction on the part of the user and the developer to fully diagnose and then fix. Within a few months, or a year, most major or common bugs have been addressed and the development team is able to move forward with completing other items on the development timetable. As new features are created, the bugs in the new feature, or the interaction of the new feature and other products, are fixed. The fewer the enhancements, the slower the pace of patches, and the less likely the new patches to the new features are widely applicable.

It is crucial to keep the originally negotiated terms and conditions in mind whenever any vendor, not just an application vendor, changes the policies involving maintenance. For example, Oracle, immediately following their acquisition of Sun Microsystems in 2010, abruptly changed its policies for both hardware and OS maintenance for prior Sun products.[3] Sun users were presented with a new set of rules requiring that they sign a new maintenance agreement for hardware maintenance for all Sun products within the enterprise or have no products on Sun maintenance. There were many other restrictions put forth. Those users who reminded Oracle of the terms and conditions of their original agreement were successful in forestalling the impact of the new policy.

Once the original license agreement is in place, and if the maintenance contract is not viable in the eyes of the licensor, then the only remaining options are to rip out the application or to outsource the hosting of the system to a more cost-effective host. It is not in the least surprising that the move toward open-systems products and hosted (cloud) offerings have been the salvation of the locked-in application user.

Moving the application from an in-house system to a hosted system does not remove license obligations if the licenses are not issued per serial number or per processor. The largest savings in moving to a hosted solution for specific applications is the dramatically reduced costs of the license and maintenance contract for the OS and the hardware, which are shared costs across multiple users.

Operating System Acquisition

Selection of an operating system is unusual, as the application requirements are operating system and hardware specific. The exceptions are open-systems platforms where users have choices between different versions of common OS, such as the Linux OS. Using the Linux base, several competitors have developed their own versions of Linux, which include enhancements and features not available in the base version, and offer their own support contracts for their versions. Unless the author offers an open domain version of the OS, there are no independent support options for any OS. As with application licenses, the OS license is protected by copyright and the terms and conditions of use are not mitigated by competition.

Operating systems are most often licensed either per serial number or per processor. Different fees are commonly charged for different types of processors, with a lower fee for the OS on a less powerful machine, and a much higher fee for the high-speed version of the same architecture. The explosion of large multiprocessor servers has given rise to per processor licensing as the price increase for the OS license from a single-processor version to a new version often made the new purchase unaffordable, which dampened sales for hardware. In cases where hardware manufacturers also license their OS, the platform (hardware) sale still tends to lead the sale in order to drag the associated licenses, custom services, and postwarranty support agreements.

In addition to the OS, there are accessory software products known as "Systems Software" that attach directly to the OS used to perform functions not included in the OS or included in a lackluster way.[4] Each of these products has its own licensing models, which do not always follow the OS license model. Before buying any new products or making major upgrades, buyers need to ascertain the price of the impact on the license costs, or required upgrades, to all products licensed with the OS, or accidently put themselves in a terrible bargaining position after the fact.

ACQUISITION MODEL: DIRECT FROM THE ORIGINAL EQUIPMENT MANUFACTURER

Businesses large enough to be dealing directly with the original equipment manufacturer (OEM) will have access to pricing and options not generally available to the smaller buyer. This does not mean that these are the best terms without investing some time to explore competitive options. Competition is essential to the sharpest pricing and best service. Competition also keeps the door open for innovation and exchange of ideas leading to better products and processes.

OEMs try to avoid competition everywhere possible in order to preserve profit margins. In highly competitive consumer markets, for example, the personal printer market, product sales may be treated as "loss leaders" for higher margin revenues from other support activities, such as ink and toner sales. In cases where OEMs may have strong competition from their own installed base, the sales force will be given strong incentives to rapidly displace older installed equipment, even if that means that the vendor captive leasing company backing the equipment sale has to lose money. Methods of distribution are lionized, used equipment purchases blocked, and lawsuits are filed by OEMs alleging copyright infringement for opening the covers of machines. All of this is done to plump the margins for the OEM.

Platforms as Monopolies

OEMs with any competition for the initial platform purchase are extremely creative in removing opportunities for a platform loss in the future when upgrades are needed. This is simply good business sense. The work needed to win the initial sale is a substantial investment for vendors and they intend to benefit from subsequent growth. Profit margins are always thinner for the highly competitive sale than for the future upgrade or for add-ons such as postwarranty service.

Creating the postpurchase monopoly is done with a combination of lock-ins that begin with the hardware itself, flow downstream into financial incentives, training and education incentives, and even operational

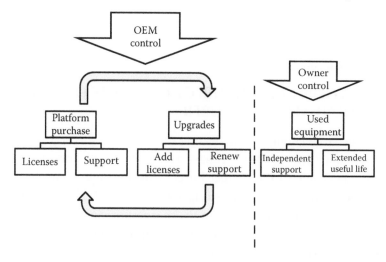

FIGURE 1.2
Hardware monopoly diagram.

lock-ins (see Figure 1.2). Before engaging in a platform decision, managers that take the time to plan ahead for future growth and upgrades will be able to control their futures when others will be completely tied to the OEM.

Multiyear Commitments

The multiyear discount package offering is common. The buyer is enticed to sign a long-term agreement covering the platform purchase, associated licenses and maintenance, and all upgrades for a period of time, often tied to the expiration of the extended product warranty. During that time, the buyer has negotiated a fixed discount level off of "list" price as in the original agreement. The discount is likely to be initially attractive compared with a single point in time offer. Buyers are attracted to this as a way to lock-in a deep discount and also avoid the effort of subsequent negotiations for the same product category. The vendor is always happy with this arrangement, which trades only a small reduction in price for a multiyear platform, services, and support monopoly. Buyers are rarely, if ever, getting a competitive discount in subsequent years.

The reason the multiyear negotiated discount is such a desirable deal for vendors is based on the normal price erosion for products as they age in their marketing cycle (see Figure 1.3). Most products are only manufactured for a limited period of time, and new versions rapidly replace either parts or whole platforms. Within 12 to 18 months of product release, used machines are generally available or totally new models with superior price

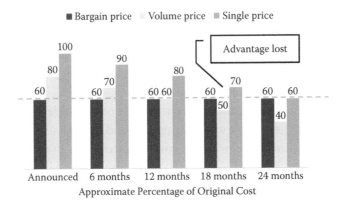

FIGURE 1.3
Package pricing over time.

performance. The deep discount of today is always higher than closeout pricing tomorrow. Buyers that have locked themselves into the multiyear contract cannot renegotiate the price point even if they are offered the new model.

Another fallacy in buyer logic is the attraction of a big discount off "list." List price for many vendors is totally arbitrary. Just as with big "sales" of 50% to 70% off list in the department stores, the vendor is not really making a special deal. The price points are set to entice the sale. True closeouts are often disposed through other distribution channels (see the section, "Gray Market") to hide the true drop in value during the product lifecycle.

Buyers that eschew the multiyear package are almost always rewarded with a lower total cost of ownership, vastly more flexibility in future selections, and far more attention from the vendor. This takes time, and some users will prefer to trade time with convenience but they will pay for the privilege later.

Within the package of attractive advantages to the buyer for these multiyear deals are special "deals" on financing, education, and maintenance. Many buyers are lulled into the sense of special treatment for these extras, when in fact all of these extras are individually negotiable and may have limited true value to the buyer.

Operating System and Machine Code License Lock-Ins

An increasingly popular tactic used by vendors to control the postpurchase environment is the tying of separate licenses (machine code or operating systems) to continued hardware support (or vice versa). Customers often think that if such practices were illegal, they could not be offered. The fact is that many of the tie-ins have not been litigated at all including antitrust actions. Most lawyers will point out that once the customer has signed the contract, no matter how ridiculous, the contract holds. Users must therefore be particularly wary of these practices, as they will not have the opportunity to make changes in the future.

Licenses for machine code and other forms of embedded software are among the most negotiable and least understood by both user and vendor sales force. Machine code, which is covered extensively in Chapter 8, was traditionally provided without a license because it was part of the hardware and thus supported in the same manner as a product failure. Over the past decade, vendors have discovered that buyers will tolerate extensive restrictions on anything labeled "code." This most likely stems from aggressive license audits on the part of applications providers. In turn, this has caused, legitimately, a strong fear of anything that might be construed as a violation of IP rights.

As a result, machine code occupies a gray area of contracting. Licensing machine code and charging for updates is perfect account control for vendors and has no advantages for buyers whatsoever. The buyer loses the ability to separate a hardware purchase from a license, and cannot continue to use the equipment without indefinite and uncontrollable license pricing and license maintenance charges from the OEM.

Machines that are combination products of hardware and application software, such as many networking and storage products, are even more subject to having intertwined and expensive postwarranty license and maintenance requirements. Many products in these categories were initially produced with these arrangements and users entered into the license agreements with full knowledge, making it nearly impossible to change vendor policies through negotiation. However, new buyers can search for opportunities to escape to vendors with more favorable terms and conditions, an exercise that is always beneficial.

Should users accept the concept that machine code (embedded software) is licensed software, they can never control the useful life of their purchases.

Users are not helpless against these tactics because of the power of the purchase order. All OEMs have to book new revenue to meet volume expectations even if margins for product sales are uncomfortably low. Most OEMs understand that they can shift margins from the product sale into longer term but more profitable maintenance and support agreements, which are rarely examined with the same intensity as the original purchase. There are many users that have capitulated before attempting to negotiate on service, assuming that whatever price pressure they bring will only push prices higher for something else. This is only true for those that do not make the attempt. It has been repeatedly proven that OEMs that face competitive pressure grant greater discounts to buyers than those that do not.

Users that withhold orders until all terms and conditions (and prices) are met have tremendous leverage that should not be wasted. Large companies are often aware of their purchasing power but are not immune to being abused.

Strategic Partnerships

One of the most pernicious arrangements for the large buyer is the "Strategic Partnership." This wording implies some kind of special relationship between the parties unavailable to others. Such "partnerships" are rarely as special as implied; it is often the case that the "strategic" element of the arrangement is a higher level of sales attention on the part of the OEM. Without the equity stake essential to a true partnership, all the negatives of the buyer–seller relationship remain intact but competition is squashed.

Any end user involved in a *strategic partnership* has put up a warning notice to all competitors that they are not welcome. The OEM is likely to become complacent on pricing and on support. Employees of the end user will be increasingly tolerant of small problems because they like or are comfortable with OEM staff. The OEM is so entrenched in the account that extrication in favor of another OEM is a multiyear process requiring significant upper management support.

Large buyers and large OEMs in these relationships tend to treat each other at the board of directors level. Pressure from the board often influences the selection of OEMs, influences the selection of leasing and banking support, and is often influential in directing consulting and "services" toward connected providers. Upsetting these relationships is difficult once established.

Competition and innovation have to be deliberately nurtured in such environments. Small or even medium-sized businesses lack the board level connections with the vendor to generate favors. Most buyers must rely upon using the options available to the vendor sales team in response to competitive pressure. The main competitive pressure point available to normal buyers is to make sure transactions are taken to the "Win Room" and at the same time presenting "Credible Competition."

The Win Room

All OEMs want to respond to competitive pressures to "win the footprint." The most common mechanism used by major OEMs is to bring highly large competitive transactions to a team of managers authorized to cut special deals for the initial sale. Both OEM direct field sales and authorized partners typically have access to this function.

The "Win Room" function is only available for winning a competitive platform battle and is not an option for repeat or upgrade business. It is therefore essential that all elements of the transaction as well as future requirements for support, upgrades, limits on increases, and so on, must be dealt with at this time. The sales team is totally interested in getting the buyer to disclose the price point at which the transaction will be awarded. Once they know the terms and prices that will win the platform, they will work their internal system to make the deal happen.

Many users mistakenly think they are getting a more competitive price by engaging with multiple business partners for the same deal. In reality, there is no advantage to engaging with multiple authorized partners for the same platform. The OEM will decide which of the many partners will be given preference in the transaction or will give everyone the same price. Users involved in a special bid battle through the win room concept need to expect to execute the deal if the target is met. Even if the agreement is "verbal," upper management of both the user and the vendor will be involved to force the deal to close so as to be "honorable." End users can gain bad reputations for themselves and for their employers and sour the relationship if this quid pro quo is not respected.

Everyone in OEM upper sales management understands this is a resource to be carefully tapped. Too many trips to the win room without a win is a black mark on the sales manager, more so than on the individual rep. If end users want to take advantage of the potential pricing of the win room process, they will have to respect the work involved as well as the politics of taking a deal up the ladder where they are highly visible to upper management. Just as users can get a black mark for *crying wolf*, so too can sales reps.

Many users are also unaware of the special relationships that exist between the OEM and various authorized resellers and distributors. Frequently partners began as former direct employees and use their friendly relationships to broker transactions within the OEM that are more favorable than can be made by strangers. The OEM sales team often dictates which partners are recommended and which are to be blacklisted. The OEM back office may also exert its influence by directing leads to favorites or by backing only those partners that have brought more business to the table than others. Unless users demand the OEM work with a favorite vendor, and do so before there is a transaction on the table, the OEM will control the partner or distribution that fulfills any orders.

Credible Competition

It used to be common practice for managers to keep coffee mugs from competitors on hand to place on their desks during discussions. The mug was meant to announce the presence of competition in the account. Not all mugs are equally threatening. Users with 100% of their equipment with one brand can set out as many mugs as they like and not be taken seriously. The only way to prove the willingness to use competitive products is to actually use some. The more diverse the equipment set, the more competitive each and every transaction is treated. This extends into the service and maintenance end of the business as complacent vendors are never as sharp as those under pressure to demonstrate high levels of performance.

ACQUISITION MODEL: CHANNEL PARTNERS AND AUTHORIZED RESELLERS

The "Channel Partner" or "Authorized Reseller" is an extension of the OEM. All of the terms and conditions of the OEM remain in place and are less negotiable through a partner than directly with the OEM. From the user perspective, the partner or reseller is an interchangeable term, describing the same relationship to the OEM. Each of these organizations is completely separate from the OEM but represents the OEM product line. The partner has many obligations to the OEM and is frequently restricted from representing any competitive lines, but in return, is buying directly from the OEM without a need to keep inventory.

As shown in Figure 1.4, the OEM distribution model does not allow for any variety or product or associated service models. The partner/channel has the same suite of options to offer customers as the direct sales force or from the retail sales channel. Options exist outside this structure but the

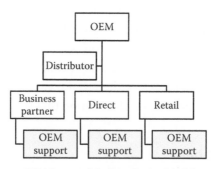

OEM Support Sales Distribution Model

FIGURE 1.4
OEM support sales distribution model.

OEM will not permit its channels to offer non-OEM options. If partners or retailers offer their own services, the OEM has either countenanced the option (in which case it is not viewed as competition), or the partner is in violation of their agreement and could be punished for offering competitive services. At some point, most partners have to completely commit to exclusively offering OEM services, or losing their distribution agreement with the OEM.

The partner/reseller channel is most often used by the OEM to support categories of business that are inefficient for the OEM to directly provide. The decision on how to deliver products and services is entirely up to the OEM, and the business model often changes abruptly. Many end users develop lasting business relationships with their partners regardless of the OEM products that the partner is authorized to resell. The user often relies upon the partner for planning, engineering, sizing, configuring, and installation of a wide variety of products, not all of which come from the OEM. Partners can be more creative and deliver a wider variety of services than the OEM alone.

Many of these partner businesses grew out of the 1980s era expansion of computing outside the mainframe and were largely applications based. For example, a lumberyard would want to automate its inventory management and billing system and would purchase a specialty software package already designed for lumberyards. The hardware on which to run the software was essentially transparent in a *turnkey* system. These applications were ideal for the equipment manufacturers to expand into these markets since the only potential buyers would be application driven (also known as "Value Added") and not a strictly hardware transaction.

Not only could the OEM support higher price points, but these relationships were ideal for the OEM since they could expand their field sales (and technical support) teams without directly hiring additional employees. Partners are still often used to manage inconvenient geographies as well as low-margin product categories. Partners can also be fired more easily than staff since ending a partner relationship does not involve employment law. Many partners have found to their dismay that they have lost their relationship in overnight and unilateral policy changes on the part of the OEM. Being a partner with an OEM can be highly lucrative but also highly unstable.

The intent of most partner programs is clearly for the partner to manage business which is not attractive for the OEM to manage directly. Therefore, OEMs also typically reserve the large accounts and most important business sectors for their direct teams. If there is a partner involved in creating the business relationship, that partner will be continuously fighting the OEM to maintain their status as the account leader.

Working with partners rather than the direct OEM sales team has advantages for the buyer, most notably more attention and flexibility. It is also more likely that a partner will work outside the strict limits of the OEM product set to deliver a more customized solution, including support and service alternatives.

Competition between Partner or Channel Sources

The partner/channel model is common as it effectively outsources the retail sales force to others and alleviates the need to hire and train (and provide benefits) for a direct sales force. Partners can be canceled far more easily than staff. For the partner, the OEM relationship is a double-edged sword. On the one hand, it provides tremendous legitimacy for the partner at the same time it eliminates the need to carry any inventory. On the other hand, the OEM can take control of any transaction directly, at any time, and the partner has no practical recourse. As a result, partners come and go rather quickly.

Partners are provided access to special pricing for reselling not only OEM hardware but also OEM hardware and software maintenance services. Partners are typically allowed to offer additional services only if the OEM does not have a competitive offering. Purchasers of equipment through the partner type program will not have any negotiating clout with respect to OEM terms and conditions, as the partner is not authorized to make any changes to the standard agreements. The only negotiation point in a partner purchase is discounting off list.

Not all partners have the same cost structure. Some of the pricing information between partners is relatively public knowledge, as with CISCO openly having different terms for platinum, gold, or silver resellers. Other OEMs are cautious, keeping their pricing a secret with less public means such as deal registration. Regardless of the standard pricing policy, there is often an appeal process the partner can use to try to get an upper hand on pricing over competitors—internal or external.

In situations where there is both an OEM account team and a partner, the OEM is often faced with having to pay commissions to both the direct sales team as well as to the partner. This is clearly undesirable as a long-term policy but is an effective way to create short-term incentives for extra cooperation between direct teams and indirect partners. The partner does not generally win in the long term unless the OEM does not want to support the client directly.

Controlling Discounts

OEMs police multiple capable channel partners in order to keep pricing high through deal registration and special discounting. In most organizations, the incumbent business partner/channel partner has priority. Commonly, whichever entity originally supplied the equipment does not just have the inside track but is also supported with a better price structure than any other value-added reseller (VAR) that comes to the table with a proposal.[5] This is not written in stone, but when businesses have to decide whom to reward, the partner that has the best proven track record in selling the OEM product is simply the most obvious choice.

Even if the incumbent partner despises the end user, the reward system still supports the incumbent partner by priority against all others. This system often means that there is no real competition between resellers, and that end users should be extremely picky about whom they choose to supply the "first" machine. Otherwise, users are forced to use specific resellers or distributors without their knowledge or approval.

Deal registration is extremely common for all vendors in managing the competitive battle between resellers. In the rare cases where there is no incumbent reseller (or the reseller has gone out of business and the account is "orphaned"), the first reseller to register a complete deal with the OEM is usually supported with better pricing. Deal registration may be perfunctory or elaborate. Many times end users are totally unaware that they have been registered by one of several bidders. Most deal registration occurs when the business opportunity is first identified and long before any formal request for proposal (RFP). Savvy end users will want to learn the OEM deal registration process ahead of entering into discussions with resellers to protect their ability to control which vendor they wish to favor.

Selecting Partners

The OEM does not always have a scientific process for recommending resellers. Most resellers started their careers with the OEM. They have long-standing friendships with their former colleagues, so when an account asks for a recommendation for a reseller, most referrals are made between friends. End users commonly expect that the OEM manages the referral business to assure end users are paired with top partners or that partners are selected equitably; however, this is not the case. Although some OEMs may actually rate or rank resellers for the benefit of end users, the referral process is most often casual. Better referrals can be had from networking with other end users and asking tough questions without first asking the OEM.

The discounting advantage to the preferred reseller is so significant that most competition is eliminated. Even when there is a potential competitive displacement (the switch from one OEM platform to a competitor), the preferred reseller is coached and supported through the win room with additional assistance and yet more pricing advantages. It is essential for end users who wish to control their choice of reseller to understand that without their direct and forceful intervention with the OEM, they will have no control in their reseller selection.

The internal relationship information and presence of any special treatment for resellers is tightly held and guarded by the OEM. As testament to the powerful forces at work, resellers who divulge this behind the scenes manipulation are chastised and threatened with removal of their reseller status.[6] As might be expected from the sensitive nature of these policies, the sales forces of both OEM and resellers are not in the loop on these policies. If asked, most reps will not be faking surprise. The best source for how partners operate is to ask the sales reps for the competitive platform. Salespeople love to take every opportunity to spread distrust about their competitor and will gleefully spill every negative point they have ever heard.

End users who understand the ways in which their choices are limited can take control if they act early enough to make demands of the OEM as a condition of doing business. Those that set up a competitive platform battle, even if desultory, will have better results than those who are just seeking a better price between resellers. The OEM does not respond to price point demands without a credible threat of losing the platform. In addition, the OEM will not support price-only battles between resellers as they will have already won the business and will have no need to discount further.

ACQUISITION MODEL: WHOLESALE DISTRIBUTION

End users are not commonly permitted to buy through wholesale distribution, but the model exists and comes into the picture as part of the system delivery, if not the direct provider. Wholesale distributors play a role in many inventory "closeout" transactions through brokers and resellers, as well as support retailers and channel partners with immediate access to inventory, configuration, and logistics services. A small number of these companies are used by OEMs as stocking facilities, effectively outsourcing their distribution and logistics.

From a purchasing perspective, there are three major problems with sourcing equipment through distribution: warranty status, gray market, and counterfeit equipment. Savvy buyers can take advantage of the pricing available through this channel without increasing risk if they respect the nuances of what a wholesaler can do.

Warranties normally flow through to the end user seamlessly, unless the warranty start date is keyed to the shipment to distribution. In these cases, end users may have less than the full warranty due to aging in the warehouse. Buyers need to decide for themselves if the remaining warranty and option for postwarranty support is adequate.

Gray Market

"Gray Market" equipment is often traded at the wholesale level. Despite the ominous description, the term refers to legitimate equipment sold by the OEM to distribution, but not delivered to an end user through the official channel. Some gray market equipment may still have full or partial warranty. Some vendors will not put gray market equipment on their postwarranty maintenance agreement as a way to drive direct sales. Gray market limitations are not based on technical problems, so independent options should be considered as part of the evaluation.

Counterfeit

Counterfeit equipment can be shipped to wholesalers appearing fraudulently as legitimate equipment. Distribution is no more, or less, a channel for counterfeit equipment than any other. (For more on counterfeit parts, see Chapter 9.) As of this writing, the best way to avoid counterfeit products is to make sure they are inspected and tested prior to delivery. This is the same process used by integrators and assemblers providing products through other channels.

ACQUISITION MODEL: RETAIL AND THE INTERNET

This is the least complex and least negotiable of all purchasing channels. Most buyers understand that if they want a "better" deal for volume purchases or customized services, they will need to work with a company better suited to the business buyer. Buyers of equipment at retail are usually small companies with off-the-shelf needs. If the OEM does not offer a multiyear postwarranty service agreement for break–fix, the retailer may offer its own plan.

Retail Service Plans

Retailers often offer service plans of their own for common consumer electronics. It is uncommon for retailers, other than a single mom-and-pop operation, to actually provide service directly. Most retail maintenance service plans are delivered by an independent repair company under contract through an insurance underwriter. The insurance company is behind the policy that is sold to the user. The company evaluates the risk of repair needs in much the same way that a life insurance company consults an actuarial table. The insurance company subcontracts to the maintenance companies and the retailer collects a fee for selling the insurance policy.

This is a common business model for postwarranty support of consumer products in general and not only consumer electronics. The model makes sense for the consumer since the repair contract is held by an entity that is probably more stable than an individual retailer. Presumably, legitimate insurers are going to be able to make good on the contract even if the retailer goes out of business. After all, a lifetime warranty is only as good as the "lifetime" of the warranty provider, not the buyer.

A considerable volume of used equipment is also sold through retail channels. Most of these products are clearly identified as "refurbished" and come with some form of warranty from the retailer. In used purchases, the retailer warranty is unlikely to be more than 90 days and 30 days is most common. In these cases, the retailer is most likely to be the actual holder of the service contract and it is reasonable to have reservations about the financial health of the retailer. Buyers should inquire which entity is backing any used warranty beyond 90 days and adjust expectations for performance (or nonperformance) accordingly.

Internet Buying

Internet purchases for new equipment "In the Box" are no different than other retail purchases, except for the usual questions of vendor legitimacy. Because continuing business relationships for repeat buying do not develop, buyers need to be careful in their specifications to make sure that they order the correct product with all the correct SKUs, at the correct engineering change level (yes, there are still differences), with or without shipping, insurance on the shipping, and so on.

The largest downside to buying by using a "Shopping Cart" and a credit card are in the area of returns. Limitations on returns include not only restocking charges, which are common, but also a limit on returns for in-the-box software in packaging that has been opened. Since most in-the-box or embedded software licenses are "deemed to be accepted" when the machine is turned on, making a return for a full refund needs to be taken into account before buying.

Using the Internet can be problematic for purchases of used equipment on several levels. First, used equipment does not have a list price and trades at whatever the buyer and seller agree, much like a used car. Second, prices posted on Web sites may be for products that are not actually in inventory, nor is the posted price likely to be the correct price without due diligence and negotiation over all of the aforementioned. Third, and classic for technology products, an individual seller may be unable to make good on any problems that arise if the equipment does not work upon arrival.

ACQUISITION MODEL: USED EQUIPMENT

Used technology equipment is a viable option for nearly every platform need, except for products in production for less than 12 months. Unless the product of interest is built for a limited market, within the first year some entity somewhere in the world will have the need to dispose of a used machine. The price point for a used machine must compete against the new purchase, so the value will drop substantially, probably 25% or more, from the original purchase price, however discounted (Figure 1.5). This is no different than driving a new car off the dealer lot.

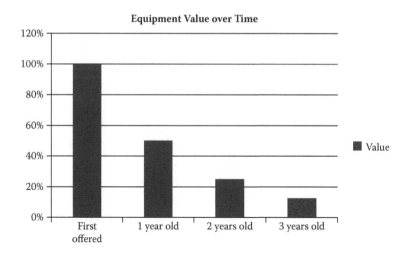

FIGURE 1.5
Equipment value over time.

Unlike automobiles, common used technology equipment value drops to single digits within 3 years, if not sooner, because of the combination of new product replacement options and aggressive vendor tactics and policies intended to prevent the used equipment transaction. Replacement products are healthy competition, but blocked used equipment transactions destroy the investment value made by users. If buyers want to depreciate their purchases, they must be able to resell them.

OEMs can block used equipment transactions in several ways, all of which can be avoided by making requirements for these actions in the initial purchase. These terms used to be common practice for major vendors through Y2K but are now at issue to the detriment of the user. There is no technical reason behind the service policy changes. The changes are strictly a marketing construct intended to drive new product sales.

1. Remaining warranty or remaining coverage under an OEM maintenance agreement should transfer automatically to a secondary user with the machine (just as with a vehicle).
2. Embedded software (machine code) should transfer automatically to a secondary user with the machine (just as the software controlling the engine control module in a vehicle transfers with title).
3. Service access codes, physical locks, and service *dongles* belong to the machine and transfer with the machine (just as a vehicle owner has the keys to the vehicle).

4. Time and materials repair must be available as an option to all owners, be reasonably priced, and include all applicable patches and fixes to embedded software (just as a vehicle owner of a used vehicle can buy service at the dealer).
5. Transfer of application and operating systems software licenses cannot be linked to the status of the machine with respect to hardware maintenance contracts (just as the owner of a vehicle does not need permission from satellite radio to transfer ownership of the vehicle).

Reasonable access to repair for equipment acquired in the secondary market and the legal transfer of embedded software that allows it to operate is essential to the smooth flow of these transactions. Wherever access to repair is limited, the value of the unit in the secondary market drops. This statement applies to everything with a technology component, from a cell phone to a mainframe.

The most desirable used products are those that the original manufacturer will optionally offer for a reasonably priced repair or support function. The used vehicle market has shown that "recertified" products with an OEM warranty command a premium price.

SUMMARY

The burden is on the buyer to fully investigate the options for long-term support in addition to OEM support before purchasing new (or used) equipment and licenses. If this step is skipped, buyers can find themselves the owners of products that

- Cannot be legally serviced beyond the manufacturer warranty
- Wrap the buyer into long-term support (maintenance) contracts without competition
- Prevent the resale of used equipment
- Prevent the purchase and deployment of used equipment
- Force the buyer into refresh purchases and license upgrades at the schedule of the vendor

NOTES

1. The concept is not unique and is well understood, particularly in software circles, where users are more frequently faced with a high volume of patching leading to new releases and a continual cycle of patching and dependence. The same is the case, but less visibly, with hardware.

2. Gartner Group created a Global IT Council for IT Maintenance in 2010. The results and recommendations for seven rights for software buyers are available at: http://www.gartner.com/newsroom/id/1403313.

3. Oracle purchased Sun Microsystems in January 2010, and in March 2010 completely changed the terms and conditions for hardware (and Solaris) maintenance. Among the many changes was a requirement that all software (Solaris) and hardware maintenance contracts be at the same level for all equipment in the enterprise, or that no software or hardware be on Oracle maintenance contracts. Many articles were posted at the time, such as "Oracle Enacts 'All or Nothing' Hardware Support Policy" by Chris Kanaracus of *CIO Magazine*, regarding the "all or nothing" policy (http://www.cio.com/article/588163/Oracle_Enacts_all_or_Nothing_Hardware_Support_Policy?page=1&taxonomyId=3045). For users accustomed to purchasing maintenance on a per serial number basis, this created havoc.

4. Computer Associates (CA) is the largest of the companies providing systems software products. Among its staple offerings are "Storage Management" products that began in the late 1970s with tape library management (now CA 1®).

5. As of 2005, the partner relationship described was the policy of IBM. It was also my experience that other vendors had similar arrangements with their partners.

6. I experienced this firsthand when as a reseller for IBM, I invited a group of clients to a breakfast "seminar" on how to control their reseller selection. Within 24 hours of sending my invitations, the senior management of my employer had been called by the CEO of IBM and was ordered to keep such information confidential. I canceled the breakfast.

2

Initial Support and Maintenance

INTRODUCTION

This chapter deals with contracting for maintenance for both hardware and software purposes in the initial negotiation. Forms of maintenance offerings are discussed, as are the common pitfalls in maintenance agreements from various types of providers.

MAINTENANCE IN THE DIGITAL WORLD

Depending on the vendor and type of equipment or software, the term "Maintenance" can mean any number of things ranging from changing toner to applying security patches to installing upgrades and enhancements. It is important for users to ascertain exactly what the vendor includes with their maintenance offerings.

Maintenance of hardware and software is needed because technology breaks and software is not perfect. Users can go without any support agreements at all, which is common for consumer-scale products, but businesses have come to rely upon technology product and application availability 24/7, 365 days a year. The more any activity requires constant uptime, the more essential the support agreement, at least in terms of emotional comfort. Even the most costly service plan does not actually prevent failure; it only prearranges for the services and parts needed to make the urgent repair.

A wide variety of major purchases are offered with postwarranty maintenance contracts in addition to the option of time and materials (T&M) repair. Vendors that do not want to be in the service business leave the business to others or engage in pass-through contracts that are executed by others. Vendors that want to be in the service business do so because they are seeking high-profit margins. Service contract margins are almost always lucrative; in many cases so lucrative that the product itself may be sold as a loss leader.

Regardless of vendor or duration, maintenance contracts are almost always prepaid in a lump sum in advance of the need for service. In this respect the maintenance contract is similar to paying for an insurance premium, although maintenance contracts are not insurance in any other sense.

The personal equivalent of a maintenance agreement for hardware or software is to have a prepaid postwarranty service contract with the vendor of one's refrigerator, large screen TV, or cell phone. Such contracts are becoming more common and are offered by the retailer at the point of purchase. Electric and gas utilities have been tapping into the market for postwarranty support of heating and air-conditioning equipment with their own prepaid or monthly billable services. At the household level, we weigh the risks (mostly price) of calling for repair without a contract against the emotional comfort of having such costs fixed. This is the insurance value of any maintenance contract.

Hardware Maintenance

Maintenance for technology hardware products means physical break–fix activities either within or outside of warranty. The maintenance agreement or warranty agreement specifies how service parts are to be supplied and how (or if) technician labor is to be provided. There are usually parameters included as to how the user should notify the warranty provider of a request for repair, and other parameters guide how quickly the provider must respond. (See Figure 2.1.)

Where there is a labor component, hardware maintenance and warranty agreements are often priced to accommodate different levels of technician response such as Next Day Response, 4-Hour Response, or 2-Hour Response. This is an extremely effective sales generator for the original equipment manufacturer (OEM) as the user is provided an "Alternative of Choice" between two to three levels of service, but never the option of no OEM service. Salespeople are trained to offer such choices and it is proven that this sales technique is consistently one of the most effective.

The vendor intends for the choice between service options to mask the option to negotiate the warranty. The entire period and coverage models

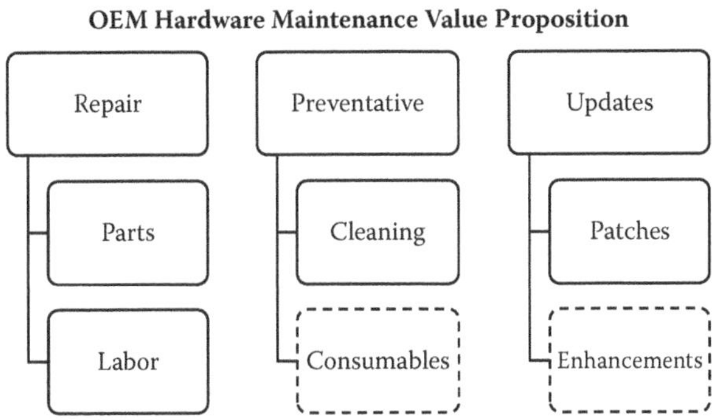

OEM Hardware Maintenance Value Proposition

FIGURE 2.1
Value proposition for hardware maintenance.

of a warranty can be negotiated. Manufacturers can back out of their costs to provide a labor contract, and they can reduce the period of warranty coverage if pressed to do so. It is quite common for buyers of large volumes of low-cost assets, such as personal computers or blade servers, to negotiate the shortest warranty possible and meet their service restoration needs with the purchase of extra units to pop in place as hot spares.

Requests to reduce warranty coverage are not voluntarily offered by OEMs because it eats into their bottom line, which should not be a concern of the buyer. The OEM sales force is unlikely to be prepared for any such requests and will not even hint that such options might be available. It is for the user to make the demand and not wait for the offer. For more on how warranties are constructed and negotiation options, see Chapter 3.

Vendors today often refer to including "Preventative Maintenance (PM)" in their service contracts. Truly effective PM was viable for analog devices but is not viable for digital electronics. There is nothing to adjust. Unless a failure occurs, there is nothing for a technician to do (short of keeping the environmental conditions within spec) to prevent failure. The most effective use of PM in the digital world is cleaning dust from fans.

True upgrades and enhancements are rarely (possibly never) part of a hardware maintenance agreement. It would not be logical for the OEM to provide a valuable new feature for no charge, nor is it possible for hardware features not already manufactured to suddenly create themselves through a downloadable patch. The parts must be there in order to be activated.

Patches provided as "upgrades" accessed within the framework of a hardware maintenance agreement are almost always repairs to defects (equipment or associated machine code) or additions to code needed to support new hardware features or models. There may be the odd situation where a new function is added to hardware, but most new functionality cannot be delivered within a patch. (Software patching can deliver new features.) Adding attachment support to a processor for a new feature, such as a new model of disk drive, is arguably an *upgrade* since the code must be developed in order to attach the physical device. The upgrade is the new feature, but the processor code must be updated to recognize the feature. The manufacturer must provide the attachment code or they cannot sell the feature.

Unfortunately, blocking access to the "updates" that support new features is a powerful weapon used to prevent used equipment from being installed at a later date. If the feature-specific code is not on the processor, then the used market for both processors and peripherals is diminished.

Users should not treat such self-serving code changes as upgrades they should pay to access. As an example, this is no different than HP producing a new model of printer and shipping a driver on a disk. The driver is only needed if the product is purchased. During initial negotiations, users have the opportunity to require all future updates and feature support as part of their purchase agreement.

Reference to upgrades within a hardware maintenance agreement is therefore suspicious marketing language and should be challenged. Buyers can flush out the real purpose of any vaguely described upgrades by asking "What is the new functionality included with this upgrade?"

Users are occasionally offered maintenance agreements that include IMAC (installations, moves, adds, and changes). Provisioning hardware upgrades and IMAC is often done by the same technicians but is not considered maintenance. These should be considered bundled agreements and dissected to make sure that all elements of the work have been properly subject to competition. For example, in an IMAC agreement that includes maintenance, the revenue for any hardware upgrades is typically booked by the hardware sales team, the revenue for IMAC falls into the revenue goals of the services team, and the hardware technician's time for the maintenance portion is billed by the maintenance team.

Software Maintenance for OS and Applications

Software maintenance is an agreement on the part of the software provider (developer) to undertake problem diagnosis and associated programming changes in the form of a patch or fix following the expiration of the initial

Software Maintenance Value Proposition

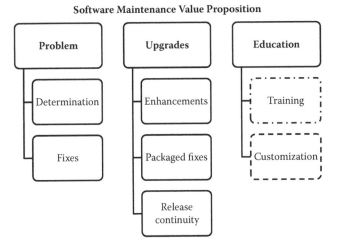

FIGURE 2.2
Value proposition for software maintenance.

license warranty. (See Figure 2.2.) A warranty is a limited time period, usually a year or less, during which time the buyer is automatically supported for these activities at no additional cost.

The type of software and the acquisition channel determines the extent and responsibility for software support. Many consumer class or small business software license agreements offer chargeable calls for "support" as a necessary way to prevent being overwhelmed with how-to questions and not actual bugs. In these situations, many honorable vendors do not charge for calls made with legitimate problems.

Within the business partner/channel partner delivery model, the partner is often tasked with handling all the inbound calls for support and being the first line to the customer for all problems, including hardware and software. Only after the partner is satisfied that they cannot support the customer from their own resources is the call passed to the OEM. Selecting a partner with important software and engineering skills for their first level support can be far more worthwhile than driving the last dollar out of the purchase agreement.

Enterprise software licenses are almost always negotiated directly with the licensor. All calls for support are almost always directed immediately to the vendor once the problem is passed through the user help desk or service desk for problem tracking and triage. The more complex the environment, the more useful it is for the user to control calls for help through a prescreening process. The help desk or service desk is usually tasked with discerning which types of problems can be solved internally and which require external support.

Diagnosing software problems is costly for any vendor, particularly when problems are not easily made to recur. The skills needed to diagnose problems and write corrections are in high demand worldwide. The software vendor does not want to be patching code, particularly code that does not generate revenue. Software maintenance is often more costly to provide than hardware maintenance because the solutions to problems have a highly trained labor component.

Postwarranty software maintenance contracts are exclusively offered by the vendor to continue the support tasks beyond the minimum real warranty for some period of years. These agreements are the cash cow of the industry and, as discussed in Chapter 1, can be extremely difficult to control.

In contrast to the hardware maintenance contract, legitimate upgrades are commonly part of a software maintenance agreement within a particular release and all associated subreleases. Typically, a software maintenance agreement does not guarantee any rights to the next release without a new license agreement. Many software vendors create upgrade "paths" to coax customers into new versions. Software maintenance agreements are highly negotiable over these points.

Consumables

Consumables-based maintenance programs are another form of maintenance largely confined to printers and copiers. These maintenance programs tend to include not only housekeeping services such as vacuuming dust from inside the frame but also reloading consumable items and changing high-wear parts on a scheduled basis. Oil changes within a warranty period for automobiles or the toner changes for a copier also fall into this category.

DEFECT SUPPORT

Defect support is at the heart of what it means to provide a warranty against defects in design, workmanship, and manufacturing. The warranty usually states very clearly the obligations that the manufacturer (or author) intends to provide at no charge as a condition of selling the equipment (or license).

Even in cases where there are no moving parts with detectable mechanical wear patterns, defects in electronics design, workmanship, and manufacturing may appear outside the initial warranty period, and be highly impactful to the user. This means that defect support is an ongoing process for both hardware and software vendors until the next release of a replacement version.

Most manufacturers announce the end of their willingness to undertake problem determination and creation of fixes with an "End of Service Life (EOSL)" announcement. The intent of an EOSL is twofold. First, to reasonably alert customers that they should not expect any further defect support and, second, and most important, to use that event as a marketing push to upgrade the customer to the next release or next model. Defect support is intentionally stopped for the older models and the release is technically "frozen."

Responsibility for defect support is the determining factor in how maintenance and support are delivered. Not all defects are minor and handled with a downloadable patch or fix. For example in 2011, Intel discovered a hardware error that required a physical replacement for their customers. This was handled without regard to warranty status, although the total cost to Intel was estimated to be around $700 million.[1] Similarly, automakers engage in costly "recalls" for products both inside and outside of warranty as their long-term obligation for defect support, including "software" recalls on such things as mistakes in the programming of the engine control module (ECM).[2]

Equipment OEMs and software developers have nursed the marketing concept that defect support should be provided under the mantle of a chargeable postwarranty maintenance agreement. This is a choice of the part of the vendor and not a technical requirement. Many vendors remain committed to providing a robust level of support as part of their value proposition for the sale. Others are enticed by the profit potential of the support agreement regardless of the cause of the underlying problem. (See Figure 2.3.)

A hardware maintenance agreement is not defect support any more than an appliance repair technician is providing defect support when replacing a broken valve. The technician is replacing a broken part with the same

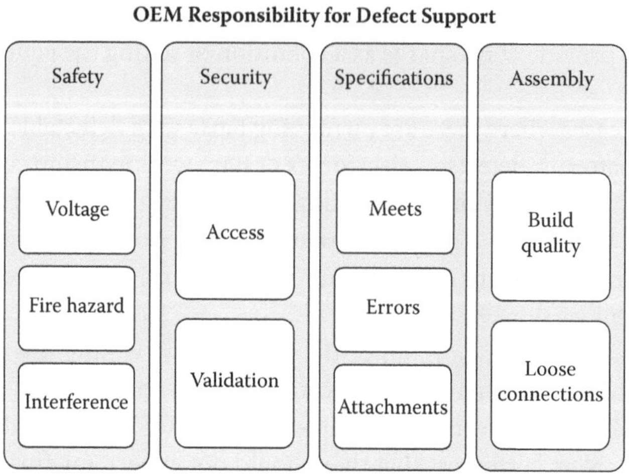

OEM Responsibility for Defect Support

Safety	Security	Specifications	Assembly
Voltage	Access	Meets	Build quality
Fire hazard		Errors	
Interference	Validation	Attachments	Loose connections

FIGURE 2.3
OEM responsibility for defect support.

part, and being paid for the labor and the cost of the part. Defect support would be if the valve manufacturer determined that the whole line of valves was defective and posted a notice on their Web site about the defect and the repair solution. Buyers of the product with the faulty valve might balk at paying to have the valve replaced, but the pricing for the replacement would not be hidden behind the fiction that the valve was not faulty and the product was being "enhanced."

Software defect support is far less easily differentiated but this does not mean that buyers should acquiesce to the notion that payment for product flaws is appropriate. Complexity of software is not an excuse for bugs; it is the reason that bugs are likely. Vendors that produce products with fewer bugs should be rewarded with more sales; however, relative "bugginess" is not yet an accessible metric.

In this context, new distinctions need to be made in contracts that illuminate how defect support is delivered for hardware and software elements outside of a maintenance agreement. Many users are being forced to purchase contracts for defect support in the guise of maintenance, which should probably be part of the manufacturer's obligation to deliver safe and functional products within (or beyond) the initial warranty.

The disruptive approach toward defect support advocated in this book is one of the most important tools remaining for equipment owners to regain control of the product lifecycle.

HARDWARE MAINTENANCE

There are two kinds of maintenance: (1) actions designed to prevent equipment failure, known as Preventative Maintenance (PM); and (2) repair in the event of equipment failure. PM no longer exists for electronic assets, but remains meaningful for printers and other output devices with mechanical parts, dust generation from paper handling, and other mechanical wear issues.

There are almost no effective early warning signs for impending failure for digital parts, so PM does not exist for digital assets in the absence of mechanical parts. (Conventional spinning disk drives are mechanical devices but still tend to fail without warning.) PM as it exists today for technology products provides housekeeping functions, such as removing dust from fans or other mechanical parts, so that equipment can run more regularly within the recommended range of operating temperatures, humidity, and airflow. Users are advised by the OEM on how to meet environmental tolerances, which is not the same thing as PM.

Most digital equipment failure is random and unpreventable.[3] With the exception of repairs made to proactively replace a part with a known problem, such as an engineering change (EC), discussed extensively in Chapter 6, most maintenance contracts are prepaid, on-call parts and labor contracts that are activated only in the event of a product or part failure. The provider of the maintenance agreement is taking an educated guess that the product is not likely to fail, and the user is betting that it will.

If users were to have the same basic information about failure rates as the OEM, they could correlate the pricing of the service contract to the need for service rather than accepting the premise of failure without any background. It is within the grasp of most users to make such evaluations of equipment that they have already purchased, but the challenge remains to have this information before making a purchase.

Repair (maintenance) of hardware in the digital world is not necessarily more complicated than the analog world, although the products themselves are enormously complex. The costs of repair are mitigated by the manufacturers' self-interest in being able to service their products without highly skilled staff. Products are designed and built to be easy to service, with parts and connectors intended to facilitate swapping components. Many products are easily serviced by the equipment owner, or could be if the instructions, tools, and parts needed were easily available.

Maintenance agreements are frequently and incorrectly viewed as a form of insurance. The owner of the equipment tends to assume that buying a maintenance agreement is a hedge against failure and the manufacturer is complicit in creating this assumption. The simple fact is that equipment fails at the same rate regardless of how much anyone paid for a service plan. The best maintenance contract, as with the legendary Maytag Repairman, is for equipment that does not need repair.

Very old hardware suffers from different types of problems that do not yield to conventional maintenance agreements. There are thousands of technology manufacturers that are out of business. Neither the parts nor the technology to reproduce the item remain. Owners of such products often cope with their operational antiques by buying spare machines anytime they appear on the used market in order to cannibalize parts. In some instances, hobbyists are able to make their own parts, the ease of which is related to the manufacturing process of the original. As might be expected, the older the model, the easier it is to see and correct problems as the parts and connections are less miniaturized.

Redundancy, Self-Healing, and Fault-Tolerant

"Redundancy," "Self-Healing," and "Fault-Tolerant" refer to features, often both hardware and software, designed to buffer users from the direct impact of equipment failure. These techniques are useful because electronics break a lot. If they did not break, networks would not need to be self-healing, and major systems would not need to be fault tolerant or fully redundant. That these marketing concepts even exist is testimony to the weaknesses of the equipment. No one should strive to seek these "features" in their equipment choices. The goal should be to deploy equipment that is so solid that equipment failure is a nonissue.

Self-healing is a networking concept behind many designs intended to avoid loss of signal when a router or other relay device is not available to do its job. This technique is effective in moving data to the destination through alternate routes, but does not address the need to repair or replace the out of service node. Eventually, the broken device has to be repaired and returned to service, which requires the attention of a technician. Users of equipment in inaccessible locations (such as those atop power poles or radio towers) are particularly sensitive to the high costs of on-site visits.

Fault-tolerant is a synonym for redundant. A fault-tolerant system has additional components intended to take over in the event of hardware failure. As more applications are developed that require 24/7 availability, more equipment is being purchased with redundant parts. In these cases, high availability is being achieved with buying increasingly redundant parts mitigating system downtime but often without improving reliability of the assets themselves.

Gaming the Service Level Agreement

The combination of remote diagnostics and call-home features on redundant devices is often used to provide end users with a technician on their doorstep with the correct parts before end users are aware of a problem. This is not the benefit that it appears. End users are being told that they have purchased highly reliable equipment, but the OEM repair process is hiding the truth of parts performance.

Without being alerted to the initial failure, the OEM could be taking days to provide the parts and schedule a technician, leaving end users without the redundant features they purchased. Sadly, many users have accepted the marketing tactic that it is a benefit to be in the dark. Without access to the remote call system, end users are being denied the ability to track and assess for themselves both the reliability of the equipment itself and the actual Service Level Agreement (SLA) compliance on the part of the OEM.

The importance of reliability of componentry is illustrated by the analogy of spare tire. Tires on automobiles are high-failure items. We therefore carry a spare tire. As we all understand from dealing with flat tires, once running on the spare tire, there is some urgency to repair or replace the spare. Similarly, any fault-tolerant system failure requires repair or replacement or the system is no longer buffered for failure.

Most data centers are paying 12% to 15% of original equipment cost (OEC) per year for maintenance services of redundant parts. It should strike end users that they could be paying less for maintenance if the parts were more reliable from the outset. Fewer parts to replace translate into fewer technician visits. Even if the parts are free, the technician is not. Selecting equipment for the most reliable equipment should be the goal of all planners but is rarely executed due to lack of measurements.

Keeping track of current maintenance events is a simple tool for evaluating the reliability of the equipment and not just compliance with SLA terms. When used methodically, automated incident tracking systems are an excellent resource for driving analysis of equipment reliability and serviceability for hardware, as well as tracking and measuring software service events. For more on this topic, see Chapter 16.

SOFTWARE MAINTENANCE

Software maintenance of digital programs in a contract sense is different from hardware in that software "glitches" occur frequently. So the first function of the maintenance agreement is to provide access to known patches to correct known problems. Maintenance or warranty agreements also obligate the vendor to undertake problem diagnosis and resolution to new problems that might occur during the contract period. In some cases, more often with "in-the-box" consumer products, the software vendor charges separately for all problem reporting and resolution.

The right to acquire fixes that were made in response to other customer problems is also tied to having a current maintenance agreement. Software releases are so commonly unstable that going without software maintenance is often a poor bargain, unless the software itself is a frozen release.[4]

Not all problems impact the organization in the same way, so software maintenance is often set up to focus on the highest severity problems (usually those that *crash* the system) as a priority. Unlike hardware, which when broken does not repair itself, most users can operate software with bugs until a patch is written and tested.

Software support and maintenance (patching and fixing) is the opposite of hardware in terms of ease of consumer or owner self-support. The authors of the software have, in most cases, unique access to their source code and associated documentation. Copyright law solidly protects the financial interest of the authors and their estate. If it had not been for the general clean-out created by the date problem of Y2K,[5] the current software support paradigm would be far worse.

The older the code, the more likely the problem that documentation (or the source code itself) is missing. Modern programming and systems design disciplines are far more demanding in terms of documentation so that the code is accessible for correction by others that did not write the original steps.

This is well understood by users trying to keep old versions of "legacy" applications in use. Not only is the source code often missing, but even when the code is still available the documentation needed to allow modern programmers to make corrections is missing. Without such essential detail, many long-standing applications are impossible to safely modify. The result is that in many older corporations, the most essential kernels of business operations remain in place behind layers of preprocessors designed around the oldest of code.

The value of software maintenance contracts diminishes greatly over time as even the most complex software is eventually stabilized and has few, if any, new patches made. Whatever "upgrades" are added usually fall into the category of adding connectivity to new peripherals (such as printer drivers) and not any new functionality. New functions that have value are almost always chargeable, either in the form of an optional upgrade or in the form of a new release.

Vendors stop maintenance on older releases for business reasons. First, software is a competitive market and new features are in constant demand, if not from end users, from the marketing team seeking to make repeat sales. Vendors cannot book new sales revenue without new products. Maintenance revenue is highly profitable, but the investors in the company expect to see sales growth for new products and new customers. Vendors thus have a strong incentive to make older versions of products obsolete as quickly as possible in order to drive upgrades and new version sales.

Vendors are also faced with high costs of development, which cannot be focused on new products if the staff expertise is still making corrections to older versions. There is no technical reason why older versions could not continue as separate supported versions other than the cost to keep staff assigned to the older version.

Copyright laws worldwide grant developers of software functionally limitless time to control support of their products. Even if copyrights were limited to 20 years from 75 years, the impact on end users would be the same. Users should not expect changes to copyright laws to protect them and have to take these issues into their contracts directly.

SERVICEABILITY

Many products are designed to be difficult to repair. This is particularly acute with consumer-grade equipment, such as cell phones and tablets. Vendors may deliberately attempt to thwart self-repair, or inadvertently thwart self-repair through manufacturing and design choices. For example, repair is extremely difficult in situations where cases are glued together or machine access requires unique tools or software passwords. Among the most powerful are warnings that tampering with the case to gain access is copyright infringement.

These types of limitations primarily serve a marketing purpose, not a technical one. Manufacturers that set out to prevent service access benefit by selling replacement products rather than extending the useful life of their designs. As we have seen with consumer powerhouse Apple, the manufacturer may also be able to sell new versions of application software, new versions of content, and new accessories (such as new power cords, adapters, and cases), all of which carry high margins and are not scrutinized for price to the same degree as the base unit. The vendor of these difficult-to-repair products also gets to control the entire repair and replacement paradigm. One has only to have owned an Apple iPhone to understand the business model.

When challenged over the legality of such policies, the argument made by Apple, and presumably others, is that they must control the "End User Experience" and are thereby justified by completely controlling repair. The U.S. Copyright Office shot down the *user experience* argument in its ruling on "jailbreaking," affirming that the owner of the equipment is in control of their experience, but the repair issue was not included in its ruling, which was otherwise specific only to Apple.

The auto industry has toyed with and backed away from serviceability limitations that are similar to those being used in other technology products, such as retail point of sale equipment and medical test equipment. In order to drive repair work to the auto dealer, owners and independent technicians were blocked from buying the diagnostic equipment and tools needed to easily identify the faulty part and replace it. The same blocks also apply to limited access to repair manuals, schematic diagrams, and, in some cases, service keys and passwords. All of these limitations were addressed in a law in Massachusetts under the automotive Right to Repair Act passed in 2012. In January of 2014 the Auto Alliance, a trade organization for the auto industry, entered into an informal agreement with a representative of the independent auto repair industry to accept the laws passed in Massachusetts nationally.[6] Unfortunately, as with the U.S. Copyright Office ruling on jailbreaking, the law was specific only to vehicles.

Buyers that require access to tools, manuals, diagnostics, service parts, and embedded software updates as part of the purchase agreement will avoid most downstream problems with controlling useful life. Once in control of all of these elements, the vendor cannot block owners from keeping the item in productive use for years beyond the arbitrary and self-serving End of Service Life and end of life pronouncements of the OEM.

DISASTER RECOVERY AND BACKUP

The Digital Millennium Copyright Act of 1998 (DMCA) includes provisions guaranteeing users the right to backup and restore copies of their licensed products for use in the event of a machine outage. These rights are set forth in Section 117 of the Copyright Code. As a result, there is little legal difficulty in porting software from machine to machine or from a backup location to a disaster recovery center, but not all contracts set up such provisions without fees or limitations. Among the most common of limitations is that of time. Most backup license language allows for only one instance of the license to be in use at any one time and puts a time limit on the number of days that a license can be run in a backup location without a formal license transfer.

There are also licenses that are "entitled" through hardware checks at start-up, which may not permit a license to boot on a machine other than the original. In planning for a disaster, the user of these licensed products needs to make arrangements to operate the software on an alternative platform, ideally without any changes in license cost.

The same antipiracy limitation and entitlement to a specific serial number of software licenses applies to media and content. Transfers of media to a different machine usually require the licensee to affirm deletion of the original version of the product from the machine as part of any transfer.

The content industry has an uneasy relationship with support, particularly anything that involves a copy of its product. The content industry is logically concerned that if the user can copy products, even for backup and restore, the temptation to keep and proliferate copies is too great. Users should expect difficulty in having flexible terms and conditions over backup and restore of these products and plan accordingly.

"APPLIANCES" AND "BLACK BOXES"

A wide variety of technology equipment is built to support only a single function. Unlike general purpose computers, these devices cannot be programmed

Black Box or "Appliance" Product Structure

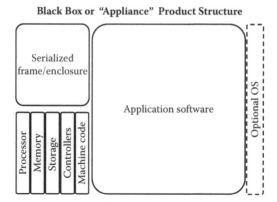

FIGURE 2.4
Black box diagram.

to perform anything other than the imbedded function (application). (See Figure 2.4.) Some examples of this category of equipment include programmable home thermostats, cable TV boxes, cell phones, encryption devices, testing equipment, and even home exercise equipment.

Although the componentry is the same as any other electronic gadget and could be salvaged for use in another device, the appliance machine as a whole is not set up to be repurposed. The programming included with the machine is not commercially licensed separately from the machine. Nor do these machines always utilize a commercial operating system such as a Window or Linux platform, which would provide a clear line between hardware and licensed software.

Support and maintenance agreements offered for this category of equipment is highly variable. Some OEMs provide defect support for the entire machine as part of the overall warranty, and others link defect support to the acquisition of separate support agreements. If an appliance owner is blocked from access to machine code patches and fixes, then the hardware elements cannot be supported separately from the proprietary software. In this case, the appliance may need to be considered a software product and licensed. This has major implications for how buyers should treat appliances for purposes of accounting and finance.

The status of the software included with the machine is critical to the rights, or lack thereof, for owners to resell equipment. If the embedded software is treated as intellectual property (IP), then the OEM may claim that the owner cannot sell the equipment inclusive of the software. In situations such as this, the equipment would be immediately reduced to scrap value, as the entire function of the machine would evaporate without the associated code.

Warranty status is also a transfer value issue. It can be the case that the machine itself will lose warranty coverage if transferred to a different location, even within the enterprise. The reasoning for this limitation is obscure and should be challenged, as many OEMs widely subcontract technicians to support their equipment in geographies where the OEM chooses not to have a direct presence. In the age of the Internet and worldwide support via toll free numbers, customers should probe carefully into the logic behind such difficult limitations.

Transfer of warranty coverage to a secondary buyer may also be thwarted through OEM policy. If the warranty transfers, then the equipment owner is usually able to easily resell the equipment and benefit from the added value of the remaining warranty. The longer the warranty, the more valuable the transfer.

Appliance products need break–fix repair just as any other. The hardware parts inside are repairable if the vendor will allow it. Appliance OEMs have tended to treat products as a single unit and do not usually differentiate hardware from software support. By associating software maintenance of their IP with hardware, appliance OEMs can prevent owners from any repair of their equipment other than their own contracts. The physical fact remains that the components are commodity items, readily repaired by any party with access to parts and diagnostics, even if the software is highly specialized. Resoldering a broken connection does not impact or violate the IP of the OEM, but if the owner does not have rights to access the case, such repairs cannot be made without risk of losing OEM support for software. This is the method by which many OEMs command exclusive hardware break–fix agreements, which could be bid competitively if the software support agreement was separate.

It has been the case that many products are so tightly controlled by the OEM that it is illogical to consider them tangible assets. If an owner cannot resell or repair an item, it is not owned and should be treated as a license. There are probably thousands of businesses that have incorrectly depreciated purchased equipment that never met the basic requirements of being an asset.

This dilemma is likely to impact more users as more appliances are deployed in formerly low-tech areas. The emerging smart grid is likely to hit many of these issues as more major appliances, literally refrigerators, hot water heaters, dishwashers, and so forth are equipped with networking and programmable capabilities. If the IP of the vendor is too tightly protected, it may be impossible for a homeowner to repair a dishwasher without violating a license agreement.

In order to avoid confusion over how to treat appliances, the following should be determined during the procurement process:

1. Require pricing breakdowns separating hardware elements from software. If there is no breakdown, ascertain the policy on asset resale. If no resale is permitted, treat the entire product as a license.
2. All software that is not provided with the machine, such as start-up code, should have a corresponding license.
3. Documentation should be provided clearly stating the rights and obligations of each party and the conditions of support.

SUMMARY

Close attention must be paid to plans for support and maintenance for the full lifecycle of equipment and licenses before entering into the initial purchase agreement. Without advance planning, the OEM agreements will feature terms and conditions favorable to the OEM and not the end user.

NOTES

1. Peter Clarke, "Intel Finds Design Error in Chip," *EE Times*, January 31, 2011, http://www.eetimes.com/electronics-news/4212699/Intel-finds-design-error-in-chip.
2. Searches for recalls based on ECMs (engine control modules) on Web sites such as DealerRater (www.dealerrater.com) reveal that many manufacturers have issued recalls for programming errors in the ECM. Such recalls are treated as ordinary recalls without charge.
3. The IEEE Reliability Society has openly discussed and debated the difficulty of visualizing a traditional "Bathtub Curve" of failure rates for digital assets due to the very short economic useful life in comparison to the known durability of the raw materials. The preference of hardware engineers is to perform extensive testing using accelerated testing methods (HALT/HASS) to try to evaluate hardware elements for component and materials failure.

4. Frozen releases are for those that the vendor has stopped all problem resolution and ceased development. Frozen releases tend to be the final iteration of the software release or production of the hardware model, and thus are more likely to be stable than prior versions.

5. Readers probably recall that older software systems were built without anticipating that a two-digit code for the year ('68 or '88) might be a problem decades in the future. Execution of routines written in the early days of computing was so tightly programmed that a four-digit code was using up valuable space in storage and memory, so a two-digit field was important. In order to avoid huge problems, which were widely and correctly anticipated, users combed through their entire portfolio of applications to modify or replace any instances of the older syntax.

6. For the entire text of the Memorandum of Understanding (MOU) between the parties, see: http://www.globalautomakers.org/sites/default/files/document/attachments/SignedR2RMOUAgreement.pdf.

3

Hardware Warranty Models

INTRODUCTION

Before selecting or negotiating a hardware warranty agreement, it is important to understand that the warranty is neither a guarantee of performance nor a free service. Choices can be made for services that are not openly offered as a standard agreement, and users can customize their relationships with original equipment manufacturers (OEMs) if aware of the purposes and limitations of warranty programs.

LIMITS AND PURPOSES OF HARDWARE WARRANTIES

All equipment breaks. End users expect the original vendor (OEM) to make repairs and replacements for a period of time under a product warranty. The length and coverage of warranties are highly variable. Warranties are not always straightforward and in many cases are far more negotiable than is commonly appreciated.

One of the poorly understood elements of warranty terms and conditions is that warranties are a marketing tool, not a delivery system for defect support. Vendors do not offer warranties without a marketing purpose; they could provide defect support for their equipment without any written requirement to do so. Vendors modify warranty terms and conditions to suit the competitive climate, offering longer warranties as enticements or differentiation.

Warranties have come to be equated, incorrectly, with product stability and quality. Although users are correct in assuming that the vendor does not want to offer a long warranty on items they know will fail, the vendor is in full control of the risk. It is a good bet on the part of OEMs that they will have minimal exposure to failure during the warranty period, which is almost never longer than 12 months for labor and 3 years for parts.

Even in the auto industry, where competitors vie for attention over warranty length, the fine print of the warranty agreement disclaims many of the likely causes of failure, such as tire wear and battery life (the "consumables" of the information technology [IT] industry), and limits their actual exposure to mechanical systems that have been time tested for decades. Engines and transmissions are not new technology and their failure profiles are well understood. It costs OEMs very little to warrant these systems. The same is true of commodity componentry in technology products. Even when new products are introduced, the first year of use is a shakeout period and most errors are discovered, reported, and fixed before the expiration of the initial warranty.

There are three basic hardware warranty models in the IT industry:

1. Parts and labor
2. Parts only
3. Depot repair

PRECONTRACTED PARTS AND LABOR WARRANTIES

Precontracted parts and labor warranties are the legacy of enterprise systems (mostly mainframes) dating back to the 1960s. These agreements were needed because the field engineer (FE), also known as a customer engineer (CE), was needed so frequently that most vendors working with large accounts had a skilled technician with an office and parts storage colocated at the customer facility. The parts were all provided by the OEM to be on full standby for the frequent need for attention.

There were options for independent support, which followed the same service delivery model in large part because IBM had been required to allow independent support under an agreement with the U.S. Department of Justice (DOJ) known as the Consent Decree of 1956. Under the decree, IBM agreed to supply service parts, allow independent technicians, and was required to sell equipment and not just rent it in order to allow for a secondary market for its products. Over time, and particularly in the last decade, the same conditions that led the DOJ to force IBM to be less monopolistic have recurred with multiple major manufacturers. Until, or unless, the DOJ or European Union (EU) intervenes to once again require open markets for products and maintenance, it is up to the buyer to demand favorable terms.

The 12-month parts and labor warranty model remains common in the enterprise/data center environment, but the trend has been for enterprise hardware systems to extend the "warranty" to a 3-year standard. (See Figure 3.1.) This is a marketing construct that successfully (from the vendor point of view) removes competition for postwarranty support, removes all negotiation of pricing for postwarranty services, and adds significant profits to the bottom line over additional years all while making customers feel that they are getting something for nothing.

The 3-year extended warranty is nothing more than a prepaid parts and labor extended service agreement wrapped into the initial purchase. The major driver for marketing this model is the lucrative nature of the postwarranty service contract for the OEM. Many in the OEM sales force do not even understand that the 3-year warranty is an uplifted or bundled agreement and will insist that 3 years is the base warranty.

The base (usually 12 months) warranty period is accounted for separately within the OEM. The reasons are entirely financial and not related to the quality of the equipment. Equipment still breaks just as frequently, but OEMs are packaging their highly profitable services into the deal to obscure and prevent comparison against independent options. It has been clearly successful since few enterprises challenge the OEM on the pricing of "free" warranty.

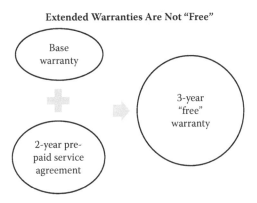

Extended Warranties Are Not "Free"

FIGURE 3.1
Building the multiyear warranty.

ACCOUNTING FOR PREPAID SERVICE AGREEMENTS

The multiyear parts and labor warranty is always a constructed prepaid service contract with revenue flowing to the service entity of the OEM. This is done so that the OEM can fully book the revenue for the equipment sale within a year. Accounting standards for public companies demand that a financial allowance be set aside to cover prospective warranty repairs. This accrual is to be set forth in financial statements. The longer the warranty, the longer the sale cannot be fully booked.

To recognize the highest percentage of revenue in the shortest period of time, the preferred method of booking sales with extended warranties is to split the deal internally as a sale with a short (actual) warranty and include the prepaid service contract as revenue to the services subsidiary. It is a thing of beauty for the OEM and a hoax for the end user.

Selling these packages of hardware and services is highly lucrative for the OEM since end users are easy prey for pitches including "free" warranty.

Even when the warranty uplift is disclosed, in a typical pitch, end users are told that the OEM is doing them a favor by providing a lump sum contract inclusive of all costs. In an additional layer of feigned convenience, the lump sum is included in a monthly lease payment. Sales representatives are trained to sell to end users that it is better for them to write a single contract and not touch it again for 3 years. "Easy" does not mean the deal is good. In fact, in the cynical world of the experienced, easy is always expensive.

90-Day Parts and Labor Warranties and 12-Month Parts Warranties

Ninety-day parts and labor warranties and 12-month parts warranties are the legacy of the explosion of midrange and distributed computing sold through business partners and other channels targeted at a specific application market. Most of these OEMs did not have a field engineering (FE) organization, so rather than invest in building an FE network, these OEMs designed products to be self-installed by the end user. Users were (and are) expected to be able to pull and replace parts on their own without specialized training.

The traditional "standard" for these products has been the 90-day parts and labor warranty where the OEM has a very short exposure to equipment failure that might require on-site labor. The largest risk of warranty exposure to the OEM is the parts warranty, which is a very small cost (electronics are cheap). OEMs offering longer warranties, such as 3 years for parts, are not adding much to their actual cost to support their equipment. Without a labor element, supporting the longer warranty is almost entirely a logistics exercise with more spent on shipping than on the parts.

Installation assistance and other technical services are typically provided by the selling business partner. Partners offering engineering and implementation services may also offer their own break–fix repair services, which largely include swapping parts. Many such agreements include warranty reimbursement to the partner for exchange of parts if they are covered by warranty.

Since postwarranty service options for these smaller systems were not originally designed to feed an in-house network of field service employees, these OEMs offered training and certification for field service (and access to parts) to their partners as well as independent service companies. Many partners develop(ed) services subsidiaries or sister companies to take advantage of the more lucrative and recurring services revenue opportunities.

OEMs have taken note of their lost opportunities to improve profits and are taking steps to build a revenue stream for their exclusive service. In these days of "flexible" labor,[1] OEMs are building their service offerings without making an investment in employees. The trend today is increasingly for authorized partners to be required to resell OEM postwarranty solutions as a condition of their authorization. There are many cases where the labor contracts are so lucrative that partners will sell product at little or no margin in order to glean the commissions for the labor contract.

Since the OEM does little except manage the inbound trouble calls, end users can make a significant savings by eliminating the middle man, in this case the OEM, and contract directly with an independent for service.

Self-Repair

Many owners, and not only consumers, have the skills and interest in supporting hardware themselves. Self-repair is only possible with some contractual cooperation from the OEM. If the product is manufactured to be deliberately inaccessible for repair, then self-repair is likely blocked in other ways, such as limited access to service parts, tools, manuals, diagnostics, and the firmware level.

OEMs make a conscious design decision to support or dissuade self-repair. The easier the repair, the less power the OEM has to drive a new sale. In many cases, it is in the best interest of the OEM to make self-repair difficult, and to make the option of hiring a skilled technician costly so as to drive replacement product sales. This is particularly common with mobile devices, such as cell phones, where the telco provides the phone for "free" in exchange for a 2-year data plan agreement. It is in the interest of the OEMs in this space to have their products last fewer than 2 years as an incentive for early replacement and early extension of the data plan.

If self-repair is an option, then all types of repair, including independent repair, are also available. It is common for OEMs to agree to ship parts and otherwise "support" the equipment owner, but to drag their feet on supplying any form of help to a designated repair specialist. In cases like this, the owner can stand in the middle to facilitate the repair and make sure to negotiate less burdensome terms in the future.

PARTS ONLY

The parts-only type of repair model came about as a result of the explosion of personal computing. OEMs needed a way to support individuals at dispersed locations at low cost. On-site labor was not practical except for the larger business user, so the parts replacement warranty came into wide use. The cost to the OEM to ship parts is modest compared with any contracts with labor commitments. Fulfillment of this form of warranty is far more of a logistics and reverse logistics specialty than a technical one.

The largest expense in the parts warranty model, for both end user and vendor, is accounting for product returns. Vendors have come to rely on the Return Merchandise Authorization (RMA) for tracking purposes. It is common practice for the vendor to invoice the user for the shipped part, and then credit the user for the return of the broken part upon receipt. This creates a hidden cost to the end user, which should be included in any consideration of total cost of ownership.

The user has to contact the vendor for the RMA number, affix the correct shipping ticket, and track the return of the part for credit against the warranty. The less costly the part, the more likely the return is not made or not made in a way that permits the credit.

Users should not consider this type of repair program as being free of charge. Although the part cost may be reimbursed, and possibly shipping, the internal costs for the service event are high. Each call to a help desk or service desk has a cost. Each escalation has a cost. Downtime has a cost, which is rarely taken into consideration but can be orders of magnitude more significant than the unit cost.

ORIGINAL EQUIPMENT MANUFACTURER DEPOT REPAIR

One of the most common repair options for low-value assets (and particularly low-weight assets) is for the OEM to designate a central repair location and have customers ship broken devices to the OEM's facility for repair. This is impractical for items that are in regular use; most organizations would not deploy assets that were not needed, so the depot repair option is often paired with an "Advance–Exchange" shipment of a replacement unit or part.

Once at the depot, the technicians test equipment for the reported fault, then either return the unit to stock if there is "no fault found" (NFF) or process the unit for repair. Completely damaged units are scrapped, but most users do not pay to ship scrap to a depot and then have to pay to have it recycled.

Users with advance–exchange programs in place will have already received an operational replacement and be up and running quickly. The dilemma for OEMs is to get the broken part back to their depot for repair. The most common method of creating the incentive for return is to charge list price for the replacement part, and then credit out the charge when the broken part arrives. This is costly for the OEM and the user, and many companies prefer to outsource these functions to a "Reverse Logistics" or "Logistics" specialist. As might be expected, logistics giants such as UPS and FedEx are major cogs in the wheel of depot repair.

As shown in Figure 3.2, users should not consider this type of repair program as being free of charge. The internal costs for the service event are high even if the part cost is not a consideration.[2]

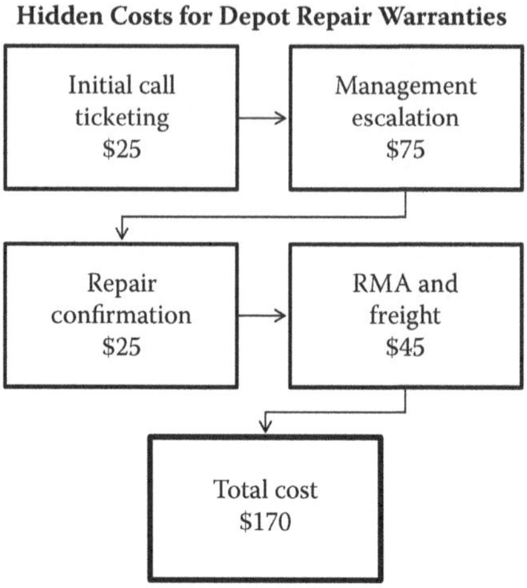

Hidden Costs for Depot Repair Warranties

```
┌─────────────────┐      ┌─────────────────┐
│  Initial call   │      │   Management    │
│   ticketing     │─────▶│   escalation    │
│      $25        │      │      $75        │
└─────────────────┘      └─────────────────┘
                                  │
         ┌────────────────────────┘
         ▼
┌─────────────────┐      ┌─────────────────┐
│     Repair      │      │    RMA and      │
│  confirmation   │─────▶│    freight      │
│      $25        │      │      $45        │
└─────────────────┘      └─────────────────┘
                                  │
              ┌───────────────────┘
              ▼
      ┌─────────────────────┐
      │     Total cost      │
      │       $170          │
      └─────────────────────┘
```

FIGURE 3.2
Hidden costs of depot repair.

LABOR

Labor is rarely offered separately by the OEM as part of a warranty program, however, independent retailers and business partners often offer their labor service as an adjunct to the OEM warranty, particularly for products that are intended for the end user to install. In these cases, technicians may or may not be officially "certified" or "authorized" by the OEM, so users should be cautious that the partner or retailer has responsibility for any errors that may arise.

Certification or authorization rules vary tremendously among OEMs. There are numerous specialists in technician labor services (also known as "body shops") that track the various qualifications needed and dispatch staff accordingly. Most independent service providers and OEMs use these labor facilitators to supplement their own direct employees.

Some vendors void the warranty if the product is touched by any party other than themselves. Although this is not common for consumer class products, it is common enough in the large IT space to warrant investigation of the ways to violate any warranty agreement before using any party other than the OEM warranty channel.

USER DEPOT REPAIR

User-based depot repair usually merges a parts warranty with a centralized repair facility. The repair location is often controlled directly by end users to support their own deployments of physically small devices in large quantities such as desktops, laptops, local printers, and increasingly mobile platforms. This model is particularly attractive to large end users needing a convenient way to support the repair of thousands of deployed devices without paying a premium for OEM break–fix labor to make individual visits to multiple locations.

Control of repair at a user depot also allows the user to control service response and service quality directly rather than through an intermediary. Coupled with a program of "hot spares" ready for immediate deployment, the repair depot has become one of the most useful service models for small items with limited configuration options. Depots are not nearly so useful for products with complex or unique software stacks and data backup and restore limitations.

Because end users can set up their own repair depots, OEMs have been cautious about authorizing such services. Policies intended to limit internal depot repair include contracts that prevent users from offering repair to other organizations, linkages of new product purchase volume to availability of parts, manuals, and limitations on the number of technicians that can be approved to provide service.

Depot repair must also be integrated with a parts replacement program to return repaired items to service as fresh spares. Most often, failed equipment generated as a result of end-user depot service is then shipped to a specialist or to the OEM for parts replacement. Most user-based depots do not have the specialized skills to repair parts, only to pull and replace parts.

Replaced parts that test NFF are often returned to stock. This happens frequently when multiple parts were replaced in a hurry to restore the unit rather than extensively diagnosed. The back office has the time to perform more extensive testing. Parts that test failed are often sent to a board-level repair specialist under contract to the OEM to recover as many units as possible. Repaired units are returned to stock as spares. Specialists in this area boast recovery rates of 95%, not including parts that are visibly destroyed, as in a fire.

In the original depot repair scenario, the OEM would certify specific end-user employees to handle warranty repairs under the contract. Almost immediately, large end users began offering their depot repair services to others, inviting the wrath of the OEM who did not want to lose yet more service revenue. There were, and still are, lawsuits filed between end users and OEMs attempting to prevent deals made with one organization from becoming competition for the OEM. One of the few cases that went to trial and had a ruling favoring the user organization over the OEM involved hospitals supporting their purchased diagnostic machines. The court ruled that the owner of the equipment could support others.[3]

Depot repair has become a common offering for OEMs and independents offering fast turnaround on high-volume products. It does not work well in practice for products where data transfer is needed, such as a laptop or desktop, because end users are unable to utilize a spare device without their unique data or settings. This is being mitigated increasingly by systems that automatically backup data allowing for the nearly immediate restoration of user details on spare devices or onto newly imaged replacements that can ship overnight.

Even when OEMs direct repairs to a depot, the depot itself is often operated by a subcontractor with a repair and storage facility authorized by the OEM. The subcontractor relationship is not obvious to the buyer. Owners can seek out direct relationships with depot repair specialists, but typically, specialists are not permitted to advertise their services to end users as part of their contract with the OEM.[4]

The depot repair process starts with the end user experiencing a problem. Most often the user is required to open a trouble ticket with the end-user help desk or service desk. Only after the help desk or service desk has attempted to restore the product with a series of remote resets and reboots and dial-ins is a service ticket escalated to a hardware repair function. If the product is covered by a depot repair contract, the vendor is contacted, and shipping instructions, usually in the form of an RMA, are provided. The end user boxes the product, attaches the shipping label, and awaits the return.

Clearly this model works best for equipment that can be easily swapped so that a replacement unit can be quickly substituted. As an aid toward addressing productivity loss, many end users supplement a depot repair program with on-site spares (*hot spares*) so that the end user does not have to wait for a replacement. Remote users are less easily served, as even a configured spare is typically a day away from delivery using overnight and express services. Some OEMs offer an "Advance Replacement" service that is logistically more complex but provides the part overnight. As expected, it is an advantage to store parts in distribution locations convenient to overnight services.

Data loss remains the largest challenge for restoring client devices. Not everyone is supported with comprehensive and timely remote backups of networked devices. Mobile devices often lose their entire phonebook (and apps) when down for repair. Other than paying close attention to backups, the best way to avoid data loss is to avoid device failure. Managers should direct their attention to failure rates (repair rates) as the single largest controllable factor in device selection and support. Do not buy (or keep) devices that break down frequently.

Assessing device reliability is not beyond the capability of any sizeable IT department. It is possible to leverage product experience as captured by the help desk and service desk to make empirical comparisons.

TIME AND MATERIALS

Not all on-site service must be associated with a prepaid service contract. Many devices are effectively repaired using on-call or time and materials (T&M) repair services. This is similar to the options available to owners of home appliances and automobiles in the postwarranty period. One can buy an extended warranty, call a local technician, or visit a local repair shop on an as-needed basis.

There are OEMs that refuse to offer a T&M option for their equipment. The reasons are always a marketing ploy to frighten customers from dropping the OEM maintenance agreement. There is no other logical explanation. No one expects a T&M repair to be free or even inexpensive. Everyone understands that per hour labor costs will be high and that service response time might be slow.

Attempts to frighten buyers over using a T&M approach to repair fall into two categories: (1) lack of priority on service calls and (2) incompetence on the part of a non-OEM repair technician.

It has always been true that a call for T&M service might not be high priority, but it is also true that multiple service options for postwarranty service are available that do not involve the OEM. Just as a refrigerator repair might not be instantly available from the first technician called, there are scores of others in the yellow pages. It is also true that an OEM technician might not be available to make the repair, but if this is the case, then the OEM may be too thinly staffed to support contracted customers as well. This is rarely the case since most OEMs have flexible labor and subcontractor options already in place to cover all service requests outside of widespread regional disasters. T&M should therefore be a viable service option for non-mission critical support.

In the case of vendors being called upon to clean up errors made by others, this is an opportunity for the OEM to look like a hero, or a revenue opportunity, or both. It is not a viable backstop for the use of Independent Service Providers (ISPs). Although it was the case decades ago that ISPs could call on the OEM to repair equipment that they could not provide either the parts or the necessary skills, it has proven to be bad business for ISPs to take a contract for equipment that they cannot support with both training and service parts.

Refusal to offer a T&M option for users does not harm the ISP, but it does harm the used market. Owners of equipment that cannot be easily repaired have tremendous difficulty selling that equipment or even redeploying their equipment into less mission-critical settings. The T&M option is often needed to restore equipment to service that has been in storage for resale or redeployment. Lack of a T&M option often dooms the product to the scrap heap.

TRANSACTION VOLUME SENSITIVE WARRANTY MODELS

Warranties that are linked to the volume of use, such as the number of lines printed or the number of transactions processed, are keeping track internally of the usage patterns of the device. The service contract, including the initial warranty, is usually set up with a threshold volume of imprints or instances of use beyond which there is a per transaction fee. Where there is competition for the initial product sale, fees per instance are negotiable as is the potential to waive all fees entirely.

For example, when multifunction printers became available they were in direct competition with copiers and faxes that were performing the same functions. The copier maintenance model was to charge "per page" or "per click" based on pages copied, but the printer model was one of paying for ink and ribbons separately from technician visits. The major differences in maintenance models became a major selling point for the multifunction printer over the traditional copier.

Many of these same products also have service (Preventative Maintenance) scheduled intervals based on pages or clicks. This is an important measurement of wear and impacts both the schedule of service and the value of the unit on the used market. In this respect, the click or page count is akin to the mileage or running hours on a vehicle. The copier that the proverbial "Little Old Lady" used in the church basement only on Sundays is more valuable as a used machine than the identical model in constant use in a busy office.

PREVENTATIVE MAINTENANCE

It is widely assumed that Preventative Maintenance (PM) is a way to prevent electronics failure. This is not the case. For electronics, not mechanical parts, service organizations no longer dispatch technicians to stare at fully functional equipment. Opening the covers and inspecting the interior does not impact reliability.

Most of what can be done to keep equipment running up to specifications comes under the heading "good housekeeping." Proper cooling needs to be maintained and often this means removing debris, such as lint, from blocking vents and clogging fans. Removing dust buildup from keyboards or gunk from trackballs is another form of housekeeping. This is not the same as PM.

Mechanical parts, such as with printer picker arms, springs, and rollers, suffer from wear and tear. These are the items where routine maintenance, usually on a manufacturer recommended schedule, will help to keep performance up to specifications. Housekeeping functions are usually performed during a service visit as most of the equipment in this category has failure-based service calls so frequently that the technician will be inside the machine making adjustments regularly.

Tape drives and tape libraries are also highly mechanical and rarely see a PM-only visit because the technician is frequently called upon to make adjustments to alignments or remove jammed media.

SCHEDULED MAINTENANCE

It is common with printer and copier products for "maintenance kits" to be sold at intervals related to wear, such as every 100,000 pages. These kits are replacement parts for high-wear items. Some maintenance kits include fresh toner, which is a consumable and not a service item, and makes evaluating this type of expense more complicated.

These types of service items are the oil change equivalent of automobile maintenance. The machines still work without the maintenance kit just as your vehicle will continue to run without an oil change. At some point performance will degrade without topping off the oil or changing the oil filter. This is the case with printer maintenance kits. Some vendor warranties include such kits, just as vehicle warranties now include routine oil changes as part of their warranty service. Postwarranty kits are typically for sale through the OEM or distributor just as postwarranty oil and filters are widely available for vehicle owners.

The owner of the equipment, if not the individual, has a vested interest in seeing that the machine is maintained in order to preserve resale value. This is why auto lessors are increasingly including routine maintenance in their lease agreements. Lessors are the equipment owners and the value of their collateral is at risk without scheduled maintenance. Banks offering car loans have the same self-interest in maintenance. The warranty holder (the manufacturer or the underlying insurance underwriter) is also interested in avoiding costly repairs during the warranty period due to neglect, which is difficult to prove.

Vendors have often created requirements to use specific brands of consumables, such as ink cartridges, in order to improve profit margins. Typically, such cartridges come with chips designed to be cross-checked with the printer to validate the brand. Manufacturers of knock-off brands of ink often find themselves in patent infringement litigation. Users are mostly warned that they will void their warranties by using non-OEM consumables. Conflicts over rights to buy and use non-OEM brands of consumables are likely to only increase with the explosion of 3-D printing. Such printing involves extensive use of resins (and other materials) that may or may not be patented proprietary formulations.[5]

SUMMARY

Buyers of hardware have more flexibility in choosing hardware maintenance options than is generally understood. The OEM sales force is highly motivated to sell equipment and can, if pressed, negotiate more favorable terms and conditions for events that might happen downstream.

NOTES

1. Flexible labor refers to technicians available for hire on a temporary or per event basis. There is a growing trend toward the use of independent (non-W2) labor throughout the service industry, including OEMs and their downstream subcontractors much as there is for other services, including accounting, secretarial, warehouse labor, and so on.
2. Per call cost estimates provided by The Advisory Council (www.tacadvisory.com) include the total cost of an inbound problem call and escalation, inclusive of labor, software licenses, and infrastructure. RMA and freight costs include labor allowances for the packing and storage of products in addition to direct shipping costs.
3. In 1998, the U.S. Department of Justice settled an antitrust action brought by hospitals against General Electric (GE) over repair restrictions of MRIs and CT scanning equipment. See "Justice Department Settles Antitrust Suit with General Electric; Eliminates Restrictive Service Equipment Agreements," July 14, 1998, http://www.justice.gov/opa/pr/1998/July/327ar.html.
4. Users can locate specialists through industry trade groups such as the Service Industry Association, see: www.servicenetwork.org.
5. For more information on the use and patent and IP challenges of 3-D printing, see Catherine Jewell, "3-D Printing and the Future of Stuff," *WIPO Magazine*, April 2013, http://www.wipo.int/wipo_magazine/en/2013/02/article_0004.html.

4

Software Warranty and Support Models

INTRODUCTION

Software support is the most difficult type of service agreement to negotiate because the developer holds most of the control. Unless there are
competitive replacement options, software vendors can, and do, exert
monopoly control over their products. The limits of warranty and postwarranty service agreements are discussed below so that negotiations are
more informed and may lead to improved contracts.

INITIAL ACQUISITION

Software warranties of some duration are included in the initial license
agreement for both operating systems and applications. The only party
that can offer a change in warranty is the licensor; ancillary partners or
distributors are not authorized to speak for the licensor, except under previously negotiated reseller or distributor rights. As a rule of thumb if there
is a discount offered by a partner or authorized reseller, the same discount
is likely available through other partners. Licensors strongly prefer for
license pricing and terms and conditions to be uniform across all purchase options. As with competition for the hardware platform, the only
time the licensor is under competitive pressure to modify warranty terms
is during the selection of a platform in the presence of competitive platform vendors.

Figure 4.1 shows the typical cycle of software support where the developer is constantly involved in fixing old problems and designing new
features, which in turn generate new problems leading to more patches

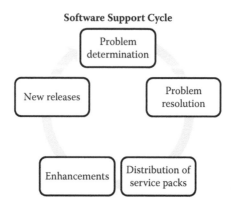

FIGURE 4.1
Software support cycle.

and fixes as the cycle regenerates. Moving away from this support model is extremely difficult because stepping outside the support agreement is intentionally designed to thwart anyone from returning to the circle.

Software Warranty Functions

The warranty itself covers problem determination and resolution within a time period and under parameters controlled by the vendor. A user must follow a formal process to report problems so that the developer can track events in an effort to discover the cause of the problem. Diagnosis of software problems (bugs) is much more difficult than hardware problems because many software issues are difficult to cause to recur. If the problem can be caused deliberately (forced), the diagnosis process is quick, although the fix may be difficult or time consuming to create.

The more complex the software, the more variable the platforms and permutations under which it might operate, the more difficult the diagnosis. The major reason that some products, such as odd brands of peripherals, are "unsupported" is that the developer has determined that the marketplace for the product is too small to support the overall costs of taking on the obligation of support during the warranty period.

Following the warranty period, users are expected to enter into a "Maintenance Agreement," which extends the time period under which problem determination and resolution is provided by the vendor. As with hardware, the license may be sold as a 3-year warranty but also be a bundled arrangement of base warranty plus a period of postwarranty maintenance uplift. It is important to know the breakdown before signing any license agreement if for no other reason than to have a proper breakdown of costs for accounting. If there is any relationship between the price of maintenance and the agreed price of the original license agreement, the price of the original agreement is particularly important to trim as much as possible so as to reduce future price increases.

In many cases, the warranty agreement also includes automatic "upgrades" to new versions and sometimes to whole new releases. Users should make sure to understand how the warranty period and subsequent maintenance periods will treat new versions and new releases, and not just patches and fixes to the current license.

Software license warranty and support contracts are often tiered for different levels of support, sometimes based on the speed of the processor, the number of concurrent users, or the number of processors. In cases where the pricing is variable, the licensor is likely to be aggressive about license audits and compliance. This should be expected as the more price concessions are made to entice the smaller user, the more tempting it is for larger users to underreport their license needs.

Different types of software have different licensing traditions, for example, the operating system for a mainframe is typically licensed per serial number and not per processor. Systems software is most often licensed by serial number and rarely by the enterprise as a whole. There are too many variations in product type based on different machines for a per location or broader-based licensing agreement to be practical. There are original equipment manufacturers (OEMs), such as Oracle, that have tried to bring a one-size-fits-all model to their operating systems and hardware support models, but they are the exception.

Contract models usually bill systems software licenses annually in advance and should include provisions for refunds in the event a licensed machine is removed. Most leasing agreements for hardware will wrap up the hardware purchase and the prepaid software license agreement into a single payment for ease of administration.

There are rare cases where specific software can be rented on a month-to-month basis with no separate requirement for a maintenance agreement since support is included in the monthly rent. This model has fallen into disuse but is coming back in the form of Software as a Service (SaaS) and cloud-based pay for usage models. Hosting services have removed much of the aggravation of supporting licensed software, which is an enormous advantage for the SMB user.

Copyright and Piracy

Developers of software are cognizant of the constant risk of being copied and their work proliferated (pirated) without being paid. This risk permeates how software is developed, distributed, audited, marketed, and supported. Software contracts for packaged products are often pages and pages of legal remedies for copyright infringement, with a few signature blocks attached. Custom projects are even more complex with elaborate statements of work, timetables, and project payments, but the same legal restrictions on copyright.

The fact is software is easy to copy. Investment decisions on the part of software developers are inextricably linked to how to protect the revenue stream. Terms and conditions (and support policies) are made to protect the revenue stream, including using support and "maintenance" conditions to turn a one-time transaction into a long-term recurring revenue machine.

Support and maintenance agreements are integral to locking the client into the recurring revenue arrangement without specifically advertising the lock-in. Maintenance, as in "Defect Support," is limited in time without an ongoing maintenance agreement, and maintenance, as in upgrades, is often included in the agreement as a further incentive. As illustrated in the beginning of Chapter 1, the software developer has an incentive to deliver products that need support in order to drive customers to continue to purchase support. If products were as stable as a rock, buyers would not willingly pay 20% or more of the original license cost to acquire patches and fixes.

This is all changing with the expansion of the "Cloud." The impact of the cloud cannot be overstated. Unlike hardware, which is being sold in great volumes to a smaller set of growing hosting facilities, the license base for traditional business software licenses is shrinking rapidly. The price performance of cloud-based services is compelling not only for small and medium businesses (SMBs), but for larger enterprises as well.

Software vendors are moving to transfer their revenue models to the cloud using SaaS and other fee-for-use models. This will in turn cause dramatic changes in how software is delivered and supported. Fortunately for copyright holders, SaaS products and billing models are far less vulnerable to piracy, as usage is much more easily monitored and measured when the software vendor is in control of the hosting.

License Audits

All software vendors are alert to the potential revenue gain by conducting license compliance audits. In the case of software licensed by serial numbers, the audit will match all machines in the known inventory with maintenance records for both hardware and software. Systems that are attached remotely will also be audited using connection tools that pull the software stack of attached machines as well as the hardware configurations and user names.

Audits for products that are licensed per seat or per instance of use are far more complicated and have created a cottage industry of audit advisors who work with clients to survive major audits. Some vendors are notably aggressive when it comes to license compliance audits and users are legitimately on edge regarding exposure to such audits. In many cases, fear of audits drives overlicensing and causes capitulation on nonlicense terms and conditions that would otherwise be rejected.

There are cases when software has been upgraded and the upgrade causes new problems. Some upgrades are so poorly designed that they must be rapidly reversed until the upgrade can be fixed. Systems software engineers dislike upgrading stable versions for this reason, strongly preferring that the upgrades be applied first by others to avoid the problems associated with being "bleeding edge" rather than "leading edge."

Upgrades and updates are also a vehicle by which the OEM introduces code that is not overtly beneficial to the end user but instead provides a new marketing advantage for the OEM. For example, a user of a high-speed printer was upgraded to apply microcode that prevented service access to the machine by independent technicians. This upgrade appeared on the machine only after the user informed the OEM that he was planning to use independent support for the machine.

MAINTENANCE PRICING

Maintenance pricing is usually based on a fixed percentage of the acquisition cost of the license. The first obvious step in limiting maintenance increases is to first reduce the base cost of the license as much as possible and then negotiate a smaller percentage and limit on future increases for the longest period possible. This is well understood and not new news.

Maintenance costs are often linked to the number of users or instances, which may dwindle over time. Users should make sure to include the requirement that they not be required to carry maintenance for dropped users/instances. Unused licenses should be allowed to transfer within the enterprise, including subsidiaries, without any additional costs and without causing new agreements with new terms.

Those users with unused perpetual licenses (also known as purchased licenses) should be allowed to resell such licenses on the secondary market as is currently legal in the European Union. (Sellers must confirm that they have deleted the presence of the product from their systems.) The situation with trade in used licenses is unclear particularly if one considers how import and export of used licenses across borders might be treated or license transfers between multinational entities with operations in multiple jurisdictions.

Users with stable systems also have the option to drop maintenance and live with the most recent valid licensed version. This is extremely practical and happens with many products in many industries and is not just an option for application systems. CISCO products are often dropped from CISCO IOS maintenance (SmartNet) because eventually products are no longer updated even when not formally announced as End of Service Life (EOSL). Keeping a maintenance contract for updates that do not appear is not money well spent. There is also an excellent underlying question of if the vendor should be responsible for providing security patches outside the parameters of an active maintenance agreement. In the absence of laws requiring such updates to be provided, users can negotiate these terms to their favor in their purchase agreements.

Vendors fight the option for dropping maintenance by using the word "security." Security is a real problem for unauthorized data or system access, but not all access security is performed or even addressed in application products. The real risk is at the operating system level. For example, Microsoft issues security bulletins and provides regular as well as urgent security updates to its operating systems, but not for its application Office suite.[1] Users therefore need to fully understand the security situation for each application product and resist the urge to react with panic over the potential loss of "security" updates for applications.

Treatment of Upgrades and New Releases

Unlike hardware maintenance, software maintenance is often sold inclusive of enhancement and upgrade functions within a release. Software vendors typically make minor tweaks and adjustments to functionality in subreleases, and the client with the maintenance agreement benefits from these features by virtue of the agreement. Many times the holder of a current maintenance agreement is granted all new releases without additional charges so long as the maintenance agreement remains in force. Lapses in maintenance are intentionally priced to discourage any drop of maintenance at any time.

Upgrades and new releases of software are often welcomed by users looking for new features. There is a downside risk to applying new code that applies to upgrades and new releases as well as patches and fixes. Changes to code are not always perfect. Despite the best of intentions and the best of testing, changes can be unexpectedly disruptive. All new versions have the potential to add problems, albeit new ones, along with new functions. It is often more difficult to remove a patch than to more fully test a patch before implementation.

Users are occasionally victims of upgrades that are not helpful. Because of the power of the word "upgrade," many users are lulled into updating software to new versions that take functionality away rather than add functions. Some vendors have been known to create upgrades that remove options for independent support—hardly an advantage for the owner. Careful evaluation of the purpose of any supposed upgrade should be undertaken to avoid painful mistakes. Only those upgrades that have been vetted by the owner should be applied. When it comes to software features the adage of "If it ain't broke, don't fix it" generally applies.

The worst case scenario for systems engineers is to apply a patch and crash the system. Most systems engineers in enterprise environments try to test patches at odd hours, on slow days, and then first on machines running less critical applications than others on which an operating system (OS) patch can be applied. Many large users have created a validation system for approving application updates and patches before any widespread distribution. Smaller users often do not have the resources or the experience to perform patch management carefully and tend to be at greater risk of patching problems, although even the largest of organizations can make mistakes.

As an example of the unintended consequences of patching, San Diego Gas and Electric (SDG&E) distributed a software patch in 2010 to roughly

4000 of its "Smart Meters." A small number (33) of the meters stopped working and cut off service. The rest stopped communicating, although they did not block service. As a result, the utility had to physically replace all 4000 units because they could not be remotely reset.[2]

PATCH MANAGEMENT

In addition to administering a healthy dose of skepticism before applying any updates, users need to keep track of all versions of patches as well as release levels of all software on all machines. This is ugly work. Software systems are available that ostensibly help keep track, but anytime there is manual input required, systems cannot include that which is not entered. Because of these expected irregularities, users often deploy remote sensing and inventory management systems to keep track of the release levels of networked equipment. Patches should not be distributed to any machine without first checking the release levels of the installed portfolio to make sure that incompatibilities are not introduced.

Patch management is a downstream consequence of having complex technology environments. From a negotiation standpoint, users cannot control the need for security and vulnerability patching; they can only control how the vendor presents such patches. There is a natural tension between the need for immediate patching to quickly close security vulnerabilities and the need to test patches before application. Users do not want to be constantly testing patches, so vendors respond by bundling patches. There is no end in sight for the battle between hackers and the hackable. Security patching is going to be with us for a long time.[3]

CONTROLLING MAINTENANCE PRICING

There are a handful of techniques that are useful for restraining the monopoly position of software vendors when it comes to software maintenance pricing. As with all other tactics discussed in this book, the best time for negotiating is before purchasing, when the vendor is most anxious to make a deal come together. Other than preventing problems ahead of time, the following are available to all users at any time.

Going Naked

Owners of stable machines with stable systems software can forego maintenance at some risk. Those that choose to drop vendor maintenance are doing so knowing that they cannot expect any problem diagnosis or resolution from the vendor, even on a time and materials basis. "Going naked" on software support is a reality for many users with obsolete systems, including products provided by vendors that have gone out of business or been acquired by hostile competitors.

Going *naked* is viable for anyone satisfied with system performance with the understanding that enhancements are off the table. Many software licenses are purchased with no intention of further changes, such as current users of Microsoft Office 2003, 2007, or other versions. Common accounting programs, such as payroll, are poor applications for going naked, as the value of the product often lies in the updated tax tables, which are included in the maintenance pricing.

Cancelling the OEM software maintenance agreement also adds the risk of losing an attractively priced upgrade path to new versions. Many software vendors try hard to keep their customers from dropping maintenance (support) and go to great lengths to keep them constantly moving to new versions.

Independent Support

Independent support of operating systems software exists but is limited in most cases due to tight vendor controls over access to source code. Owners of the copyright usually defend their copyright to such code zealously. A third party (also known as independent) cannot modify code if they do not have legal access to the source code. There are versions of some proprietary code, such as MPE/iX from HP, where the source code has been licensed to select independents to support, but even so these independents are limited in what they can do. They can write binary patches to specific problems but cannot write a new version of the OS for resale, such is the case with the many versions of Linux that are written and supported independently. Linux is a truly open platform that is not supported at all by the creators of Linux but is instead used as a base platform for others to enhance and market as their own versions.[4]

Even with proprietary systems, users often use independent consultants or agents to help them navigate the support needs of the product. This includes handling the work required to manage patches and updates as provided by the software company. Most software maintenance agreements have limits on the types of handholding that might be desired by an end user so that agents of the owner can fill the need without violating copyright law.

Switching Systems Software Vendors

There are few options available to end users for competitive replacement of systems software, particularly operating systems. The OS is designed to run on a machine-specific instruction set provided by the hardware manufacturer. Although many hardware products are designed to allow virtual machines to operate emulating other products, the basic OS must still provide the interface.

Users have come to accept that the choice of hardware platform also dictates the systems software suite. Even though the software and maintenance/support services are priced separately, extricating oneself from a platform is rarely simple. The most common method for removing a system that has become too costly to support is to outsource or outright replace the applications rather than attempt a push-pull switch, often dubbed a "heart transplant." Once the application is ported to a different platform, the link between the hardware vendor and the application is broken and the undesirable platform can be removed.

Applications are far more difficult to switch. Most applications are integral to the basic operation of the organization, and even when available for multiple platforms, may require a significant commitment to training for use on a different system. Replacing an application entirely with a competitive product is a major undertaking. Conversions to new systems can take years and cripple business operations for months. The expected end result of major application changes may never materialize.

Vendors of major application systems understand their importance to the licensee and have been known to take advantage of their powerful position to dominate other agreements. As an example, after Oracle purchased Sun Microsystems in 2010, it abruptly altered the terms and conditions for support and maintenance for Solaris as well as for the Sun and StorageTek hardware platforms. Users would have fled the Sun/Solaris platform had they not also been entangled with Oracle as a major application vendor.[5]

SUMMARY

Users should proactively consider the end of life of software products as part of the initial acquisition. Consideration needs to be made for how licenses might transfer or be dropped from maintenance before buying any licenses.

NOTES

1. For a useful discussion on the risks of continuing to run an operating system past EOSL, see a Microsoft article regarding the end of support for Windows XP in April 2014. Tim Rains, "The Risk of Running Windows XP after Support Ends April 2014," *Microsoft Security Blog*, August 15, 2013, http://blogs.technet.com/b/security/archive/2013/08/15/the-risk-of-running-windows-xp-after-support-ends.aspx.

2. Onell R. Soto, "Faulty SDG&E Smart Meters Replaced," *U-T San Diego*, May 21, 2010, http://www.utsandiego.com/news/2010/May/21/faulty-smart-meters-replaced/. The story was widely covered in the smart grid and utility industries although the problem and the product set was software.

3. For an excellent and accessible article on security patching, see the National Institute of Standards and Technology (NIST) "Guide to Enterprise Patch Management Technologies" draft, September 2012, http://csrc.nist.gov/publications/drafts/800-40/draft-sp800-40rev3.pdf.

4. See Allegro Consultants (www.allegro.com) for further information about MPE/iX support and Red Hat (www.redhat.com) for more details about their supported versions of Linux.

5. I was part of a team of independent maintenance providers that met with the U.S. Department of Justice over perceived antitrust violations on the part of Oracle in the summer of 2012. Prior to that meeting, we had many discussions with end users who were infuriated by the policy changes, but none were willing to come forward, even under nondisclosure agreement, to affirm their position based on the fear that Oracle would use their involvement as a reason to find them out of compliance on a license audit. This was a rational fear. In 2012, Oracle posted roughly $1 billion in revenues from its license audit activities.

5

Finance and Accounting Issues for Maintenance and Support

INTRODUCTION

A correct and accurate breakdown of acquisition costs between hardware, software, and services is at the heart of proper finance and accounting. Only hardware can be capitalized and become part of the assets of the organization. Hardware depreciates over time, and tax benefits and taxes are based on the value of the hardware. Anything other than hardware cannot be reported in this manner. This chapter is about the inadequate and peculiar treatment of modern technology assets lacking accurate ownership details.

INITIAL PURCHASE

A wide variety of products with technology components are sold in some combination of hardware, software, implementation services, and warranty, and then financed or leased. When these elements are blended together into a single lease payment, it can be extremely difficult to ascertain how much of the payment is associated with depreciable and tangible assets, and how much is really just financing of intangible and nontransferable licenses and prepaid services.

Vendors are often reluctant to provide the breakdown of "soft" and "hard" costs because such information makes it easier for the buyer to shop around for alternatives. To avoid being shopped for price, sales reps are trained (brainwashed) to insist that the customer does not want such information and they are doing the buyer a favor by keeping everything bundled into a single price. This is illogical, and if taken at face value is entirely to the advantage of the vendor. The details of how purchase contracts are built are essential for both proper accounting and for administration of the finance or lease contract.

Technology equipment buyers are often faced with differentiating between which parts of the purchase are tangible assets (depreciable hardware) and which are intangible (software, installation services, maintenance contracts, wiring, custom programming, etc.). Businesses are particularly sensitive to the accounting treatment of tangible assets for issues of taxation (personal property tax), depreciation (income tax), company value (stockholders), insurance (replacement value), and even liability. Even when buyers choose to treat a machine as 100% software for depreciation purposes, the presence of a hardware frame proves that some form of combination is involved.

Many products, not just "Black Box" products, are sold without a clear breakdown of tangible and intangible elements. Some values, when presented, are preposterous and should be challenged. For example, a machine sold for $1 million inclusive of 3 years of on-site hardware and software support cannot possibly be 100% tangible. The breakdown must include reasonable allowances for elements that are licensed and not transferrable (such as machine code, OS, or applications) and must also take into account the projected costs of labor and parts to support the prepaid service agreement (often mischaracterized as a warranty), which is also intangible.

Ancillary services related to the installation, including wiring, installation of environmental systems such as air conditioning, fans, moisture sensors, and so on must also be separated from the technology purchase, and either expensed appropriately or depreciated as part of the improvements to the facility.

The real danger is that if the machine cannot be resold as a functional machine (such as inclusive of key licenses), then the real value of the machine may be scrap not just at the end of the warranty but within the warranty period. If there is no secondary market during the warranty period, then the lender is at a great financial risk if it needs to repossess the equipment due to default. If there is no secondary market value for the asset beyond the warranty period, then any investor in a lease cannot expect to get a return on equity based on either a renewal or a resale in the secondary market. Thus, the treatment of support for both tangible and intangible elements is integral to finance and accounting.

A number of lessors and lenders have turned a blind eye to these issues because they have business incentives to write more leases and loans. Poor attention to residual risk is not a problem for the end user, who is the borrower, but investors in these instruments are involved in more risk than they generally understand.

Residual value is critical to the leasing and lending of technology equipment. Used equipment that retains its value is necessary for users and lenders, but a double-edged sword for the original equipment manufacturer (OEM). The manufacturer has to sell more equipment to meet its quarterly sales projections, even if the equipment being replaced is still being depreciated by the owner or a leasing company. Many OEMs have had a series of leasing companies backing their transactions, each of which has dropped the line of business because they were always being prevented from writing renewal business.

Lessors are also appropriately gun-shy about transactions involving equipment with no secondary market. If the equipment cannot be resold, no one can recoup their investment if the lessee stops paying. The more an OEM crushes its own residual values, the fewer lessors will bid for lease business. Eventually, the only option becomes the OEM itself, which should be a huge red flag warning to buyers that the vendor is going to dominate the useful life of the asset beyond the initial warranty.

TANGIBLE ASSETS AND INTANGIBLE SOFTWARE

Technology hardware is largely assumed to be tangible property. Owners of technology equipment expect to be able to capitalize the purchase and take depreciation over time. Owners of hardware can theoretically borrow against the value of the asset and the total value of capital assets is part of the financial strength of the organization. The breakdown of tangible assets (hardware) from intangible licenses (software) is fundamental to financial accounting (Figure 5.1).

Licenses, for both operating systems (such as Windows) and application products (such as Excel) are well understood to be intellectual property (IP) and treated correctly under accounting rules. Other types of intangibles that are commonly a part of a technology hardware purchase, which must be expensed under generally accepted accounting principles (GAAP) rules, include installation, consulting, "services," and prepaid maintenance

Tangible and Intangible Parts of Systems

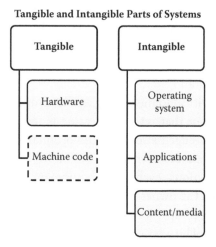

FIGURE 5.1
Tangible versus intangible elements.

agreements for hardware or software, which often require a little digging to pull out of bundled proposals.

The lines between tangible and intangible have been blurred deliberately by many vendors, largely because users no longer ask. These distinctions now require user engagement. In many cases, the OEM no longer provides a breakdown of values and must be pressured. Even more challenging is dealing with the status of the embedded software that comes with the machine but might not be specifically licensed. Since end users do not know such software exists, nor its role in accounting, users are vulnerable to being denied important use of their purchases over such hidden details. The potential for abuse over embedded software is so large that all of Chapter 8 is devoted to the topic.

Accounting for machine code has not caught up to the peculiar distortions that are being brought into the marketplace by manufacturers over intellectual property rights to machine code.

In most cases, machine code is not mentioned in any equipment purchase agreements. Machine code is an integral part of all electronics and so long as there are no problems, such code goes unnoticed for the life of the asset. Often, machine code simply arrives with the asset and transfers as part of the asset. Products as varied as industrial process control systems to automobiles to major appliances all incorporate machine code to drive key functions if not all functions.

Treatment of machine code for purposes of fixed asset accounting is the lynchpin between tangible and intangible value. If machine code travels with the machine without additional expense, then the equipment purchase can be treated as a capital asset and depreciated. Leasing companies can invest in the equipment because it has residual value, and banks will lend using the equipment value as collateral.

If, on the other hand, machine code is treated as intellectual property by the OEM, then the machine cannot be transferred (sold, or used as collateral) without paying license fees to the OEM. Licenses are not capitalized and must be expensed. Leasing companies and lenders do not have the security of the asset value as collateral and thus charge higher rates for lending, if lending is even an option. During periods of tight credit, or with organizations without stellar credit, many lenders will not support financing of licenses (intangibles).

LENDING AND LEASING

Under a lease or a loan, the lessor or lender owns the equipment that is used by the lessee. The lessee does not totally own the equipment until such time as all payments are made (as in a capital lease). Many leases require that the equipment be returned to the lessor under the terms of the lease. Lenders and lessors are naturally concerned that equipment be properly maintained in order to protect their collateral interest in the equipment. They are also concerned that their equipment has a secondary market outlet other than the OEM, otherwise they may be stuck with equipment that is repossessed due to default.

Users should note that if there are not any independent lending or leasing options available for equipment under consideration, there is almost certainly no secondary market for the item.

Lack of a secondary market for specific equipment impacts not only the willingness of lenders and lessors to risk money at favorable rates; it also means that terms and conditions may be unfavorable to treating the equipment as an asset from an accounting perspective. Asset-backed financing is based on a combination of the credit quality of the borrower (lessee) and consideration for the value of the underlying asset. There are many borrowers that could not acquire the equipment needed to operate their business based on credit quality alone. The stricter the lending standards in times of difficulty, the more difficult it is for smaller or newer businesses to get loans of any sort. Asset-backed financing is often the only way that such companies can acquire equipment for any terms other than cash.

Lenders and lessors in every industry evaluate the product and product vendor carefully in order to be confident that the asset that is protecting their investment is going to hold its value. If the vendor is known to prevent resale of equipment, the lender will take this into account and price the lease accordingly. The higher the asset value over time, and the more easily the equipment can be resold in the secondary market, the higher the risk the lender will take on the asset. If the asset is difficult or impossible to resell, the lender will treat the transaction as a pure credit risk. Leases or loans based strictly on credit are almost always priced at a premium, if they are offered at all. For small and medium businesses, it is not uncommon for personal guarantees to be required to make such transactions possible.

Intangibles, such as software licenses, postwarranty maintenance contracts, installation, wiring, and consulting, are not normally leased. Lenders and lessors may accommodate a percentage of such costs within the framework of a larger tangible asset acquisition, but despite the accommodation, the underlying loan or lease rate is higher to account for the lack of collateral. Every lessor and lender does this, even if the monthly lease payment to the customer appears as a lump sum. If a sales rep insists that the rate is all-inclusive, then he or she did not have anything to do with structuring the details of the transaction.

Lenders typically accommodate lending against intangibles only when there is a high percentage of tangible value associated with the hardware. Otherwise they may, if they offer the option at all, lend at rates appropriate to an unsecured line of credit or require a personal guarantee. It is therefore important that all buyers make sure that they buy equipment that can be resold so that it can form the basis of an asset-backed financial transaction and not be paying a premium for treating the same product as unsecured credit.

Users are best served by avoiding policies that reduce asset value before they make any agreements to purchase or to lease. There are already many ordinary limitations on value in the secondary market, such as age of the equipment and the popularity of the item to secondary buyers. Even for those buyers that have no intention of keeping any leased items beyond the end of term, the viability (or lack) of the used equipment market underpins the lease rate as shown above.

Buyers that intend to capitalize equipment have the same interest as lenders in making sure their equipment investments have residual value in the real market and not just an arbitrary percentage of depreciation. If the equipment owner, for whatever reason, wants to dispose of assets, use assets as collateral in a general business loan, or transfer equipment to sister entities, the assets must have value or the original transaction was incorrectly booked. Vendors that sell equipment as capital equipment have an obligation to assure their buyers that the equipment can be resold and to provide the mechanism for such transactions to occur.

The following list reviews common technology equipment policy practices that destroy secondary markets. Some of the practices are extremely common and have been in place for decades, which does not mean that they are good policies for the end user.

1. Contracts that allow postpurchase changes to the agreement in the future
2. Hardware that cannot be sold as a functional machine
3. Hardware platforms that are not separable from the application ("appliances")
4. Hardware platforms that require additional licenses to operate (operating system [OS] or application)
5. OEM service policies that will not support used equipment
6. OEM service policies that do not offer time and materials repair
7. OEM service policies that are not serial-number specific
8. OEM service policies with return to service penalties

INTERNAL ACCOUNTING

The Financial Accounting Standards Board (FASB) allows owners of tangible assets to capitalize their purchases and report the value of the asset as part of the value of the company. The values are expected to decline over time and are subject to depreciation schedules mandated by law. Thus, the value of the company is linked to the value of the installed assets.

Companies use the value of their assets as collateral in business loans. The total value of the company includes the value of its entire base of owned equipment according to the depreciation schedule. Equipment that is being depreciated is presumed to have real value if it were to be liquidated.

Technology products are particularly tricky to capitalize because many items cannot be easily resold.

FASB rules also provide for treatment of rental and leasing agreements as distinct from purchases. Some of the current rules have special status in accounting practice that can be advantageous financially to business. For example, "Operating" leases allow payments to be expensed in the same manner as consumables and office supplies. Operating leases currently require that the lessor have at least a 10% residual risk in the transaction, that there not be any bargain purchase agreements at the end of term (which would remove the risk), and that the lease not be for greater than 75% of the economic life of the asset. These criteria were made in 1976 when technology equipment had very long useful lives and high residual values.

Operating leases are increasingly difficult to qualify than in prior years because technology equipment loses value rapidly for reasons of application obsolescence. Matching the lease term to the useful life of equipment in order to qualify for operating lease treatment is now challenging. Many types of equipment hold so little residual value at the end of just a few years that operating leases, under the rules set by FASB-13 (the applicable section of the code) are impossible without some "tweaks."

Such *tweaks* require the agreement of the end user to achieve. Lessors will offer many suggestions, and it is up to the end users and their accounting and legal teams to determine if any are acceptable. One legitimate and common arrangement is to reduce the lease term to match the useful life and residual risk parameters, but very short lease terms are often impractical because the user does not want to have to return or negotiate renewals in very short periods. Other less straightforward options include bundling the equipment lease into an overall "services" transaction, but such arrangements are difficult to structure without the cooperation of the vendor providing both the services and the equipment.

The treatment of these off-balance sheet lease transactions have been the subject of nearly a decade of discussion within the FASB and are close to being eliminated in favor of on-the-books treatment of all leases.[1] Under the proposed changes, users will be required to treat operating leases more as capital leases, so that the financial obligations be more visible to investors. This should increase the scrutiny being placed on the real value of assets as distinct from licenses.

Users that fail to scrutinize their purchases carefully for the real value of assets before putting them on the books can be inadvertently committing fraud under Sarbanes-Oxley.[2] There are other places where financial or tax fraud can send users to jail. For example, one cannot lease or rent equipment and be allowed to sell it; they are not the owner. There is an entire section of the Uniform Commercial Code (UCC) that consists of public filings by lenders and lessors announcing their security interest in equipment that might otherwise appear to belong to the end user. Unscrupulous end users have been known to sell or trade in equipment they do not own. Unscrupulous lessors have been known to repeatedly sell their interest in leases to multiple parties at the same time.

Manufacturer (Original Equipment Manufacturer) Captive Leasing

Manufacturer-sponsored leasing programs have been around since the beginning of the computer industry. They remain among the most popular of options for several perceived reasons:

1. The OEM makes the transaction simple.
2. The OEM can provide a true operating lease (under current FASB-13).[3]
3. The OEM appears to offer more favorable terms and conditions.

As with everything else discussed in this book, all is not as it appears.

Ease of Contracting

The OEM captive leasing program may be competitive, but the user cannot know this without first seeking competitive prices. At the very worst, a buyer looking for competitive financing and leasing options validates that the OEM has provided a competitive offer. The buyer is never harmed by introducing competition into both the product purchase and the financing. Buyers can choose to skip this step for their own reasons, but they are not being thorough.

Novice end users are unlikely to know that all lessors, including the OEM captive, use and need the same types of paperwork. Any lessor can claim they "make leasing easy." The problem for novices is that there are lessors with very one-sided leases, including many famous OEMs. Making a lease easy does not correspond to contractual fairness. Many of the "gotchas" common in technology leases are discussed later in the chapter under the section "End of Lease Issues."

Leases are much more complicated than purchases. The lessor will have control over everything other than the use of the equipment. The lessor, as the equipment owner, has the right to demand the equipment be kept in good working order and will almost certainly demand that all maintenance be done by the OEM service team, regardless of expense or quality. The OEM will require, as do all lessors, that the equipment be covered by insurance. Lawyers that specialize in leasing should review all lease agreements to protect the user from unpleasant surprises. In short, the OEM contract is not any different in terms of ease than any other.

Operating Lease Treatment

The OEM can provide the fiction of an operating lease because it controls the pricing of the equipment to the end user. Many buyers do not realize that behind the appearance of the residual risk requirement of the operating lease, the OEM is discounting the cost of the equipment to the leasing entity behind the scenes. The lessor (usually the captive) is not at the same, or any, residual risk because the lease is being directly subsidized by the OEM. Even when the lessor is not the OEM captive, the same behind-the-scenes discounting is done by OEMs in order to support the client's desire for operating lease treatment. Buyers with no intention of leasing may have additional price leverage since the net to the OEM for a cash purchase is roughly the same as the net through a leasing entity, once the hidden discount is taken into account.

Partners and other channel providers can typically offer the same leasing terms as with the OEM directly, on behalf of the OEM, or with independents. Buyers are more likely to have independents involved in partner- or channel-provided relationships than direct with the OEM because the partner has more flexibility in choice of lessors if they choose to use it. OEMs know that independent competition exists, so they have a compensation plan to their partners that rewards them for moving the lease transaction to the OEM captive.

On the third point, many end users believe extra attention to problem resolution comes with an OEM captive lease. This is a fallacy. The leasing arm is always a separate business from both the sales organization and the support organization and has no involvement—zero, nada, zilch—in the service delivery process. The sales team may beg for more help to keep a key account, but the lessor is not involved even as the OEM fights to win new sales. Leases are written to keep the lessor out of the middle when it comes to equipment satisfaction; the term "hell or high water" means that no matter how dissatisfied the customer, the lease is to be paid.

The rate at which equipment needs repair is also entirely independent of the amount paid for the equipment, any promises made by the sales team, and the financial institution involved in the lease or loan. Equipment breaks down at rates determined by the quality of the design and manufacturing influenced by the environmental conditions of the end-user site.

With only a handful of exceptions, which change too frequently to make any lists meaningful, leases are always written to be sold off to other lessors or discounted by lenders for the income stream. Leases can be packaged into investment vehicles and resold in batches in financial markets. There are a few organizations, such as IBM, that have made a specialty of leasing and holding their leases using their extremely low cost of borrowing, but even so the IBM lease agreement is intentionally written such that it can be sold if it wished to do so. The terms and conditions within virtually all OEM leases are similar to one another so that any lease can be bought or sold as the parent company determines.

Lessees may think they are dealing with the OEM entity during the lease term, but they are always dealing with a separate entity from the moment the lease is signed. The lessor (including the OEM captive) purchases the equipment from the OEM under an "Assignment of Rights" agreement. The assignment agreement transfers all the written terms and conditions that were made as part of the negotiation to the buyer on behalf of the lessee. The lessor, even when owned by the OEM, is set up as a separate company to allow the manufacturer entity to book the sale of the equipment once the lessor pays the manufacturer.

The manufacturer does not want to have any overhanging obligations, called "Contingent Liabilities," which limit their ability to recognize all revenue from the sale. As explained in Chapter 3, the vendor has to account for the potential cost of delivering on the warranty agreement in their financial statements. Other types of liabilities, such as performance of systems, are often requested and just as often denied for the same reason. Any situations under which the customer might want their money back, or a reduction in the lease payment, or any other changes that impact the rights of the equipment owner (the lessor), make the lease agreement impossible to sell.

Service agreements, which often include performance metrics, are particularly difficult to lease for this reason. Lessors, even the captive lessors, are not set up to be in the services business. If the user has the option to cease making payments in whole or in part, the lessor loses the ability to sell the lease, or even discount the lease to a lender. Lessors work assiduously to avoid these entanglements. The most likely arrangement within a bundled service agreement is that the kernel of payments due belonging to the equipment remain a "hell or high water" obligation and the services provider can never waive those terms. The OEM does have the option to guarantee the payments to the lessor or guarantee the lessor a buyback in the case of performance issues, but the lessor is always made whole or there is no lease. Lessors, for their part, can make business decisions to take a chance that the OEM will perform and work with a lender that will hold the lessor responsible in the event of a dispute. These types of lending arrangements are called "Partial" or "Full" Recourse based on the responsibility of the lessor in backing the transaction. Lessors hate to write such deals, because they tie up their own capital in risky transactions. Manufacturers and their leasing captives are therefore in a unique position when it comes to supporting such complicated transactions. The manufacturer can shift revenue and risk internally in order to make a deal happen.

At the end of the lease, the owner of the lease, including the OEM captive, is always in competition at the end of term against the new equipment sale. The lessor prefers to keep the leased unit in place and write a renewal or sale in place in order to meet their profit objectives. Captives have the same goal, but those that have retained ownership of the equipment can be compensated by the parent company to give up financial gain in order to help win the new sale. Similarly, captives can be caused to waive billing for the almost inevitable problems that come with making an equipment return.

Users that do not have any current new business under discussion with the OEM have no bargaining power at the end of lease even when the OEM owns the lease. Similarly, problems with leases written with independents cannot be favorably negotiated at the end of term without some enticement, or agreeing to new leases for replacement equipment. On the whole, with the exception of the few vendors that hold their own leases, the assumption that the lessor and the OEM are working together at the end of term is false.

Residual Value

It should come as no surprise that historically OEMs with terms and conditions intended to facilitate the secondary market also carry the highest residual values. This in turn creates more incentive for end users to purchase the product with the longest lifecycle. Financial institutions in turn support the acquisition of these brands because their investments are better protected. Leasing of automobiles and other major assets follow the same pattern. Lessors offer lower rates for products with higher residual values than those that are scrap at the end of term.

In the absence of residual value, the equipment owner should be taking the fastest depreciation available by law.

The secondary market is essential to supporting equipment value over time. Lenders offer lower rate loans where there is collateral value than for services or software that lacks residual value. The soft cost elements of products that are intangible are not transferrable collateral in the view of lenders and lessors. The more an IT device is treated as software, the less it is an asset. Rates for loans are higher, if they are available at all, where the percentage of software or other tangibles are higher than when the loan is for 100% hardware.

Licensed Software and Residual Value

Licensed software support terms and conditions can also kill equipment residual value. If license renewal or license maintenance costs are made high for machines coming to the end of term, the end user will face a fabricated situation intended to make a new purchase more attractive than keeping the older equipment. The OEM wins a new sale, but the user never gets control of the asset and cannot "sweat" their assets.

A case in point is the disk array. Providers of these products have created financial packages that bundle all hardware, software, and associated maintenance into a single monthly payment. At the end of the term, the renewal or maintenance costs for the license portion alone is made deliberately unattractive so that the proposal for the total monthly cost of the replacement product is always more attractive. In this way, using software license pricing, the OEM has created the situation where there is no residual value for these products. In turn, because the OEM has not only crushed its own residual value, lenders will not take the equipment as security in the transaction. The OEM wins not only the constant churning of more equipment sales but also controls all financing, trade-in, and the secondary market.

It is true that the hardware array itself can be broken down into parts for secondary market use, but no whole machines are traded as a consequence of OEM license transfers or renewal fees. OEM lease terms and conditions are set to prohibit filling up empty arrays with used disk drives, so the only market for used equipment is the very few parts that can be absorbed by independent repair companies.

End users frustrated with being manipulated in this manner can use the power of their purchase orders and cease dealing with companies that engage in these practices, or start making their own contractual demands for more favorable terms and conditions. There are always options, including operating on frozen releases and independent service.

The treatment of all forms of license transfer in the information technology (IT) transaction has the power to support hardware value or remove it. The inability to transfer basic machine code (essential embedded software) with a machine, despite the lack of a license agreement supporting the restriction, is a deal killer for used equipment. Buyers in the secondary market will be wary of any equipment they cannot operate without either being in violation of IP or are unable to have serviced by the OEM.

TETHERED AND "BLACK BOX" PRODUCTS

Accounting treatment of equipment known variously as "appliances," "black boxes," and other "tethered" devices, is particularly challenging. Typically, such products are prepackaged for a particular function, such as a network sniffer, encryption device, or other specialized gadget, inclusive of all hardware and specialized software. They are sold as whole machines with little or no option or requirements for end-user involvement. Typically, the device is attached to the system and runs. If there are configurable details, the OEM provides the configuration service as part of the installation.

Most of these machines are built from parts that could be used to support other applications but are not sold as general-purpose machines. All such products have a hardware backbone with all the embedded internal software associated with the parts, upon which the specialty application software rides. (There may or may not be a separate operating system layer.)

Because the machine and the application are inseparable, the vendor often refuses to provide a tangible–intangible breakdown. Lacking a breakdown, end users run the risk of miscategorizing licenses as hardware, in which case they will have misstated their financial position, or being forced to treat the entire expenditure as a license.

Users of appliance-type products are often prevented from using any break–fix or service options other than from the OEM. This may not be obvious in the initial proposal. Clues that such options are not available begin with the pricing. If the proposed product is not provided with a clear breakdown of hardware (tangible) and software (intangible) elements, it should be a red flag to the buyer that separation of service and support may not be permitted. The license in a frame should never be treated as an asset, whereas the frame with a license can be broken down as at least partially depreciated.

The following questions should help determine if the machine should be considered a license in a frame or a frame with a license.

1. What is the hardware value of the machine for accounting purposes? If the machine has no hardware value for accounting purposes, it is not a tangible asset despite its physical appearance and should be treated as a license.
2. With respect to the hardware components, how much of this package includes break–fix support? If the machine has hardware value, the portion of the value assigned to any service contract must be expensed and not depreciated.
3. What are the terms and conditions under which I can transfer this machine to another user? If there are limitations on the transfer of the machine as a working unit separate from the application license, then the machine may have no value on the secondary market, in which case it should not be treated as an asset.

There are also many cases of bundled hardware/software solutions where the values assigned to the hardware elements are irrational. A hardware configuration based on commodity PC parts will carry the residual value of the same parts used as a PC. Using such parts to drive a sophisticated application does not increase the value of the parts even if the vendor greatly overstates the pricing. If the parts are commodity elements then the pricing for the portion maintenance service agreement should be comparable as well for similar agreements for similar parts, without extra value assigned to the specialization of the software.

The same caveats apply to customized versions of common platforms resold under various turnkey application programs. For example, there was a popular turnkey application sold to automobile dealers based on a modified IBM AS/400 (iSeries) platform. During the auto industry bailout and subsequent reorganizations, a number of dealers went bankrupt and the machines came onto the used market. The machine itself was not a standard IBM model and was in a frame with the logo of the application vendor. Despite being a fully functional machine, the only value of the machine in the used market came from the standard IBM-branded parts.

END OF LEASE ISSUES

There are many ways for lessors to make money at the end of operating leases. Users should take steps in the negotiation process to protect themselves from at least the following, all of which are directly applicable to repair and maintenance.

Lessors traditionally make stipulations regarding equipment condition when equipment is to be returned. Every item being returned will be evaluated closely for necessary repairs and exterior appearance. Users should take care to have all of their leased assets in full working order with all original configurations intact before attempting a return. If changes have been made, such as upgrades, any parts that were added should be removed, the original parts restored (and tested), as all other additions will become the property of the lessor.

Scratched, dented, or worn items will be at issue over the wording "ordinary wear and tear accepted." This phrase is common boilerplate, which if incorporated into a lease agreement is inadequate without specific definitions appropriate for the product. The fact is that secondary market buyers will not purchase machines that appear used and lessors cannot sell equipment that no one will buy. The lessor will not tolerate the loss of value due to user mishandling and will charge the lessee accordingly.

Lessors want their assets returned in the most valuable condition possible and this includes microcode and firmware updates. Users should expect to be required to keep their equipment up to "current EC (engineering change)" levels as part of the lease or finance agreement, even if the user does not need the updates. The language in most leases is slightly archaic since the "EC level" reference is no longer sufficient. Equipment can be up to current ECs and still lack important microcode and firmware updates if such updates are not described as ECs by the OEM.[4]

In the case of deployments involving large quantities of similar assets, users will want to make sure they have rights to substitute serial numbers for like products at time of return. This can be messy if not set forth properly in the agreement. Over time, many repairs will have been made using unit swaps rather than on-site repairs. Keeping portfolio equipment lists current is rarely done well and is costly to accomplish. In order to avoid the seemingly pointless burden of tracking serial numbers, many users will assume they can return other equipment instead. In practice this is difficult since:

1. The original product may no longer be available in order to substitute.
2. Configurations of the original acquisition may have changed.
3. Ownership of the substitute item may be unclear.

Lessors can, and often do, take the opportunity to make the substitution of assets extremely painful and costly.

Lessors may require that all equipment in the lease be under an active maintenance agreement. If the equipment lease and the warranty period align exactly, this is not an issue. However, if at any point users wish to avail themselves of self-repair or independent repair, the lessor may have rights to approve the change. Lessors typically permit non-OEM maintenance so long as there are no added restrictions on resale that would flow from such actions. Unfortunately, there are now many OEM policies that directly limit options for independent maintenance and these should be negotiated out of the purchase and maintenance agreements before the lease is put in place.

Maintenance contract stipulations pass beyond the original lease term and are incorporated into any renewal period, so users should be careful to avoid limitations on repair by creating the original lease agreement with adequate flexibility.

SUMMARY

Buyers must make sure to require a complete breakdown of hardware and nonhardware elements prior to purchase. Hardware elements should be fully depreciable, repairable, and have resale value independent of any software or services. Anything that cannot be resold, or can only be resold with the permission of the OEM, should not be treated as a tangible asset.

NOTES

1. See proposed changes to GAAP accounting from the FASB at: http://www.fasb.org/jsp/FASB/FASBContent_C/ProjectUpdatePage&cid=900000011123.

2. Sarbanes-Oxley provides for criminal prosecution of executives that knowingly sign off on financial accounting reporting (for public companies) that they know to be incorrect. Sarbanes-Oxley requirements are particularly important when it comes to evaluating any form of "tweaks" to lease agreements to cause them to qualify as operating leases under FASB-13. The same exposure will apply to putting assets on the books that are not assets. Although the financial obligation of a lease including intangible elements may be correct in terms of payments, the company cannot treat "air" as an asset and be in compliance.

3. The FASB standards for operating leases are undergoing extensive change as of this writing. It appears likely that many former "operating" lease criteria will change to require on-the-books treatment of these formerly off-the-books transactions. The impact on the OEM captive as a lessor will be marginal since they traditionally had little or no financial risk in the transaction behind the scenes.

4. ECs, safety, security, and driver updates, as distinct from other forms of microcode and firmware, are not often defined in user purchase or maintenance documents. This is an area of tremendous conflict in the arena of access to third-party mainte-nance in general and not just for lease returns. Users should consult their attorney for better phrasing of the rights they should request.

Section II

Postwarranty Support and Maintenance

6

Responsibility for Defect Support

INTRODUCTION

Understanding which parts of the original equipment manufacturer (OEM) organization (Figure 6.1) are responsible for which types of defects allows end users to craft more equitable and productive service agreements. The sales force from the OEM is unlikely to be prepared to delve into this issue and may easily provide misleading or incomplete details. Nor are all OEMs straightforward with buyers as to which types of defects are covered regardless of warranty status, such as "recalls" in the auto industry, or those that are only available within the postwarranty maintenance agreement as a separate contract.

SOFTWARE DEFECT SUPPORT

Experience tells us that the vast majority of calls for help from users to help desks are not hardware failures or software failures. Actual hardware problems are a tiny fraction, certainly less than 10%, of all calls for help.[1] Of the remaining 90% of calls, at least 40% are for user problems such as settings or passwords. That leaves 50% for software problems that are tricky to diagnose and more difficult to fix.

One of the first, and most important, jobs of the help desk (or other support team) is to determine which type of problem is involved and then direct the call to the most appropriate entity. Many times the user needs help with how to use an application or has a setting issue, in which case the problem is not a failure issue and is handled without involving the vendor. A large number of companies subcontract for these types of

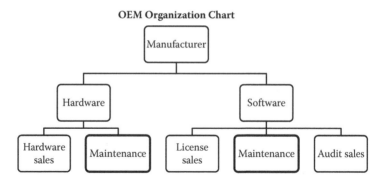

FIGURE 6.1
Manufacturer organizational chart.

how-to problems, including investments in education, Web-based learning, and so forth.

The next level of work involves distinguishing between hardware problems and software problems. Hardware problems are usually simple to identify because the part in question does not work and cannot be made to work even intermittently. Machines with broken connections do not fix themselves. There are no "fault tolerant" circuits, only functional or nonfunctional circuits. In this respect hardware failure is totally binary, meaning "on" or "off."

Software failures are typically intermittent. A wide variety of conditions may have to come together at the same time to cause a software failure, which is why the reboot function is so frequently effective. The reboot returns the machine to the original settings, which clears the error. It is axiomatic that the more applications in use at the same time and the more interfaces that are active, the greater the chances of problems of interoperability and outright conflict.

The most common way to start diagnosing a hardware failure compared to a software problem is to restart the machine. If the machine restarts correctly, the problem is software. The word "failure" is more appropriate at this stage because in the world of hardware, failures reflect something broken, which must be the case, whereas a software problem can have a workaround or be ignored if the consequences are not significant. Really horrible software failures, such as those that repeatedly crash the machine or the active partition, are simpler to diagnose than others and usually get the highest level of attention (for severity) from vendors.

That being said, there are a few nuances of software use that impact hardware failures. First, if there is a failure of machine code and patching resolves the problem, this is still technically a hardware failure as the responsibility for such patching lies with the hardware OEM. These are rare events in practice. Most patching is done preemptively by the OEM in the earliest days of initial warranty and otherwise only reloaded if code is lost during a repair. Engineering change (EC) level options also fall into this category.

Second, there is some validity to overclocking a processor contributing to heat-related failures. Overclocking is usually a hobbyist and gaming technique for faster performance. In this respect, the use of the software does cause hardware failures, but the repair remains a physical one. One cannot reset the clock speed of a fried processor and return it to service.

Similarly, there are situations where high demands on read-write activity on disk drives contribute to drive failure as the excessive activity is presumed to accelerate the failure rate of actuator arms through overuse. Again, this is ultimately a hardware failure even when the causation is clearly a problem that is controllable by software. The actuator arms will not repair themselves, so the technician must still place "warm hands" on the device to make the repair.

Once hardware failures are ruled out, the team must then attempt to differentiate between operating system and application system failures. Without belaboring the obvious, in a situation where multiple applications are run, if the machine operates correctly with some applications, but not with one specific application, the source of the problem can be easily narrowed and the vendor contacted. Operating system bugs are more likely to be systemic and impact the system regardless of which applications are running.

Depending on if the application or the operating system (OS) is causing a problem, the responsibility for defect support for those problems lies with the software vendor. Determining which vendor needs to fix which elements of code is a contentious area where there is a high potential for finger-pointing. There is some validity to the idea of reducing the level of finger-pointing by dealing with a single vendor, but there is still the matter of problem resolution.

Users are often attracted to the idea of "one throat to choke" in the context of problems in general. It is assumed that by having one point of contact for one vendor, that problems are resolved more quickly. Under the convenience of the single point of contact lies the unpleasant truth that many vendors have grown by acquisition and not by developing a truly integrated set of products. One has only to look at the acquisition history of software giant Computer Associates (CA) to see that completely disparate products are marketed together but were never built to be installed together.[2] Similarly, many hardware vendors have purchased both hardware and software vendors that were never built as integrated products and packaged the products together.

Types of Patches

Patches to hardware and software problems are often distributed as "urgent," "security," or "essential" updates in addition to those patches that add support for new features or drivers. There are often a lot of patches, and patch management is an important task for system administrators to track.

Since downloading patches occasionally introduces new problems, many organizations test all patches before applying the patches more widely. The axiom "If it ain't broke, don't fix it" is widely applicable in the world of software maintenance. Many serious outages have been linked to patching without adequate testing.

The wording of "Security Patches" is particularly suspect for all application software as it is very much to the marketing advantage of vendors to create a sense of urgent need for continued maintenance contracts for their products. So long as "Cybersecurity" remains a fear factor, vendors will use wording that implies that dropping software maintenance is going to make systems more vulnerable to attack. This needs to be carefully scrutinized before being accepted. As with all types of marketing hype, the buyer needs to consider the motives of the source.

HARDWARE DEFECT SUPPORT

In the technology equipment industry, hardware break–fix and defect support for underlying machine code is often blended into an extra-charge service agreement typically sold as "maintenance" or "extended warranty." These agreements cover both patches to machine code as well as parts failure, they are actually two different types of problems resolved by two different groups within the OEM organization.

Diagnosing flaws in machine code errors as distinct from manufacturing or component issues is the result of analysis of recurring problems that resulted in service calls. The call for repair under warranty is the first indication to the OEM that there might be a hardware problem. It is not possible to determine if the cause is a logic flaw or a manufacturing flaw until the part (or machine) is returned for examination.

Forensic analysis (autopsy) on returned parts is done to ascertain the root cause of the failure by reliability and quality assurance engineers in the back office.[3] In addition to the OEM, each part manufacturer has its own staff specializing in root cause analysis. If defects in design can be addressed with a downloadable patch, an update is issued. If the defects are more profound, it is likely that the item will be replaced by a new model. The new model is commonly identified by an "engineering change," or EC level.

Defect support of logic flaws (machine code) is one of the major tipping points within service contracts determining if the owner of the equipment has any opportunities for self-service or independent service. If defect support of machine code is provided by the hardware OEM as part of the hardware, it is common for the end user to be able to select from multiple support options. If on the other hand, defect support for machine code is bundled into software maintenance agreements, then it will be unlikely that the owner will have any hardware support flexibility.

Logic Flaws (Hardware or Machine Code)

At the most basic level, computers are combinations of logic statements expressed in wires. It does not matter if the wires are large and visible, or extremely small. Logic is always part of the design of the computer, and logic errors cannot be corrected externally, unless manufactured into the chip. Defect support of logic must therefore be provided by the manufacturer. Such errors are almost always corrected without cost to the buyer because the OEM would not likely be able to sell flawed products once the flaws were identified.

Logic problems exist in both physical form, as manufactured into the chip, and also in the embedded software that comes with the machine. Embedded software has many names, machine code or embedded code being the most dominant but also including microcode, firmware, BIOS (basic internal operating system), PLC (programmable logic code), and IOS (internal operating system). The function of the machine code software is twofold. First, machine code provides common routines needed to perform strictly hardware functions such as moving data from storage and back. Second, machine code is also a more flexible delivery system

Hardware to Software to Hardware

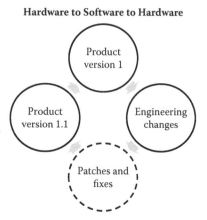

FIGURE 6.2
Hardware to software to hardware.

for corrections to complex logic that can be more easily corrected in the event of problems. As shown in Figure 6.2, corrections delivered dynamically, as is often described as a "Firmware Update," are only one of many media options. In many cases, patches to machine code are collected and included in subsequent versions of the hardware itself, thus completing the cycle of machine code back into a manufactured form.

Within the OEM organization, fixes to hardware errors and machine code errors are supported by the hardware designers, uniquely capable of finding and patching errors in logic. This is an entirely different set of employees from developers of operating systems or application systems that design their products based on the specifications of the hardware. Corrections from the hardware engineers may be distributed in physical form in a replacement part manufactured to incorporate the changes or in a less costly form as a distribution of a patch to machine code. The net result is identical. The part or machine patched using media is identical to the part or machine later manufactured inclusive of the microcode corrections.

OEMs strongly prefer to avoid physical recalls and if at all possible distribute "patches, fixes, updates, and so on" to be applied in the field. The method of distribution involves either some form of media or more often a download to media from the library of patches using the Internet. In the early stages of a product release, there are always more patches than later as field use reveals patchable problems. This is why a series of patches may be recommended at the time of equipment installation even though the product is "new." This is the same concept familiar to users of the Windows line of operating systems, known as "service packs."

There is truth to the phrase "bleeding edge" instead of "leading edge." Significantly new products will expose new problems in the initial release. Buyers of new products should expect to invest more time in patching. Some products are more widely distributed in the marketplace than others, so models with less common distribution are less likely to have a thorough "shakeout" in a short period of time.

Eventually, the majority of problems are identified and patched. At this point, the machine code stabilizes and few new patches are needed. Most patching in later years is done to allow native mode attachments of new peripherals or to recognize new features or major model changes. There is always the potential that a functional patch might be created years into product distribution, but this is rare since the development team for the hardware is almost certainly working on newer generations of equipment.

Support for older models is eventually dropped officially at the "End of Service Life (EOSL)" or "End of Life (EOL)." This should not be confused with end of functional life. Many products remain in productive use decades past their last EOL or EOSL announcement.

Keeping equipment in service for the long term requires that equipment owners have access to the full library of known patches for several reasons:

1. Not all users want to apply patches that are not immediately needed.
2. Machines or parts in storage will not have had all patches applied until they are deployed.
3. Machines do not work correctly without all updated code.

Without access to the library of patches (or patch media), the owner is unable to assure him- or herself that their equipment is operating correctly at any point in time, including years beyond the initial warranty.

Updates versus Upgrades

Manufacturers have been taking the position that they are entitled to limit access to updates to machine code because they are constantly providing new enhancements and should be paid for their work. The flaw in this argument is that the enhancements they reference are unlikely to be the types of improvements users expect when they pay for upgrades.

Upgrades are defined in the computer science realm as:

1. A software program that provides added enhancements over an earlier version.
2. A hardware device that provides greater performance than an earlier model.[4]

Updates to code intended to repair flaws (patches, fixes, security patches) are not improvements. It is only the confusion of end users that allows such word play to be tolerated. Updates to machine code are not upgrades. Updates are synonymous with patches and are the *recalls* of the technology industry.

Upgrades to functionality come with component purchases. No machine code *update* will change the physics of the product. If one wants a faster processor, they must buy a faster processor. If there were a way to double the speed or density of a product through a machine code upgrade, the wise OEM would charge for such value and not hide it in a patch.

Defect Support of Component Failures

Parts suffer physical breakdown in addition to logic flaws. Different batches of parts may perform differently. Different assembly teams may do a better job than others. Just as automobile buyers used to joke about never buying a vehicle built on a Monday morning (presumably the workers were recovering from hangovers), every step in the assembly process introduces variables. Shipping and handling exposes parts to extremes of heat, cold, vibration, and static charges. Given the variety of ways that delicate electronics can be abused prior to installation, it is quite an achievement that we have such low levels of immediate failure (known as "infant mortality").

Almost all infant mortality is observed in the first attempt to power on the machine. If the machine powers up and executes all self-diagnostics, the machine is no more likely to fail than any other. Loose connections may reveal themselves as machines are transferred from the staging location to the end user, but there is no need for a "Burn-In" period, which is an artifact of mechanical production or of warming up the device, which is an artifact of vacuum tube technology. A 90-day warranty is more than enough time to identify and exchange any shipping or packaging defects.

Once a machine is installed, the likelihood of specific part failure will track that of the overall mean time between failure of that part. In a configured machine, the overall failure rate of the machine will be no better than the worst part. For example, a CISCO switch with some components rated at 300,000 hours (clock oscillator) and others at 40,000 hours (fan) will still need service at the rate of the fan regardless of the rest of the parts.

OEM support of parts failure is a key element of all hardware maintenance contracts. Issues of infant mortality are usually resolved with an apology and a quick shipment of additional units. All other parts issues are resolvable with parts replacement. The choice of labor for parts replacement properly belongs with the equipment owner, just as owners of automobiles and refrigerators are able to choose if they prefer to pull and replace parts themselves or hire a technician.

The supply chain for parts is large and complex. It is possible for defective parts to be knowingly sold and integrated into machines. Some parts defects may have a low potential impact on users and be integrated into machines where the impact is unlikely. Others can clearly be unscrupulously sold and enter machines where the flaws will cause problems. It is up to the quality assurance team at the OEM to carefully test and differentiate between offerings by competitive parts manufacturers.

Last, defects in electronic hardware may be problems with machine code and not just physically broken parts. Defect diagnosis and repair is an obligation of the manufacturer, and, although presented as a "benefit" to the end user, it is incorrectly positioned. The end user would strongly prefer to purchase equipment that had no defects. In most markets outside of information technology (IT), defect support follows the machine through the chain of owners and does not disappear by location. This is the case for automobiles, major appliances, hot water heaters, and televisions. It should also apply to technology products of all kinds.

ENGINEERING CHANGE LEVELS

There are occasionally hardware product problems so large that a part needs a physical adaptation to operate correctly. These types of repairs are usually identified early in the product cycle and managed by the OEM under warranty with an engineering change (EC). ECs are physically different parts. (Wherever possible, lower level parts can be updated through microcode updates to machine code that can be applied dynamically on installed equipment.) Whether updated in the field through a code update or physically replaced, the EC level is important to the correct functioning of the machine.

Keeping track of the EC level of parts is burdensome but must be done. EC changes fall short of the types of problems that would result in a wholesale product recall and may not impact all customers. Unfortunately, owners of equipment without the latest ECs on their equipment are also losing value since the secondary market buyer will almost certainly demand the highest level EC as part of the purchase. End users who have equipment with available ECs should make sure their OEMs update all their equipment as soon as possible to retain value.

ECs can only be provided by the OEM and are product specific. Because such changes are costly for the OEM to provide the labor for the repair, the trend in the industry is to move away from a physical EC and design products to be updated through remote delivery of patches to machine code. Remote delivery, be it through the Internet or using physical media, is still far less costly to the OEM than manufacturing and replacing parts.

Manufacturers have successfully avoided being chastised for delivering buggy and unstable products largely by describing patches and fixes as updates or, worse, upgrades. True enhancements or upgrades to equipment are almost always offered as chargeable offerings and not included for free in any EC or machine code updates.

REMOTE DIAGNOSTICS

Many devices in large enterprise settings are equipped with remote sensors and communications links to report failures to the OEM without the user having to make a trouble call. Many end users believe that these systems are constantly monitoring the health of the device, when in fact they can only sense failure after the fact and not prevent failure. For example, large disk arrays are assembled using massively redundant devices with sophisticated mirroring software so that individual and common failures of disk drives do not interfere with the work being done. These systems cannot detect future failures, they can only report on actual failures. The call home feature merely reports the failure so that a replacement part and technician can be efficiently dispatched.

Why do such machines not know about impending failure? Because there is no way to know until a circuit fails. OEMs would know more than anyone about the mean time between failure (MTBF) and the mean time to failure (MTTF) of the device, yet it is entirely clear that they do not have such information or they would build and deliver much less troublesome devices. It is vastly more expensive for an OEM to dispatch a technician to replace a $40 disk drive than to have avoided the problem with a $.10 difference in unit cost. This, more than any marketing claims, tells us that the OEM does not yet know enough to predict failure.

Remote hardware diagnostic routines are provided by the OEM to reduce the volume of physical service calls needed for diagnostic purposes. It is clearly efficient for the service provider to have the ability to triage a problem remotely in order to either restore the unit to service remotely (as with a reboot), or to manage the logistics of providing both spare parts and a suitably trained technician to the location. This is the reasoning behind the common disclaimer in Service Level Agreements (SLAs) that the response time to the SLA begins only once diagnostics have confirmed the hardware problem, not the instance that the service call was placed.

There are several problems with reliance on the remote call-home function. First, the remote service may not be linked to the owner's trouble tracking system, making the management analysis of repairs made and calls for service extremely difficult. Second, the remote call-home feature is not always the trigger to the acceptance of the service request as criteria for response within the SLA. Many times the vendor will treat the inbound call as preliminary and not start the SLA "clock" until some other validating step is performed. This allows the OEM a significant time jump on arranging for technicians and parts ahead of the user's knowledge with the result that the SLA can appear in compliance. The user can thus be made more exposed to concurrent failure far more extensively than is known.

Remote fixes are not the same thing as remote diagnostics and mostly refer to the option for a technician, usually a software specialist, to connect through a network into the computer and remotely manipulate settings. There is no such thing as an actual remote repair, since anything physically broken must be physically repaired.

There are new forms of remote diagnostics being used for service delivery in many industries, such as a telematics function in the auto industry.

The telematics connections between the onboard computers of the vehicle and the service department of the auto dealership are already the subject of concern by privacy advocates. It is not yet clear if the data being exported to the dealer belongs to the user or the dealer. Privacy advocates prefer that the owner of the equipment also be the owner of whatever data is transmitted, for the obvious reason that it is truly no concern of the dealer how the vehicle is being used (unless the dealer is also the owner as in a rental agreement). Users expecting to buy equipment with telematics functions should consider for themselves if issues of data privacy are meaningful and negotiate accordingly.

There is already the potential for the OEM to block access to the telematics function by owners with expired warranty agreements. It may come to pass that telematics services are sold separately from traditional hardware break–fix, in which case the owner of the vehicle (or other device) may find themselves required to purchase a telematics license in order to have dealer service. Although this type of tying agreement sounds illegal, it is exactly the same as many current requirements in the IT industry with different names. Buyers wishing to avoid contracts that bind them in the future should reject such options, if necessary, in order to protect their future rights to service.

SUMMARY

Obfuscation by OEMs is common when it comes to the details of defect support. Buyers should be digging deeply into all proposals to make sure their contracts correctly represent their understanding of how defect support is to be delivered and at what cost. Many OEM policies are surprisingly negotiable on the matter of defect support, and every opportunity should be taken before any purchases are made to demand the most appropriate agreements possible.

NOTES

1. Analysis done by TekTrakker for clients revealed this general pattern of problem calls to the user help desk/service desk. Readers can easily cull their own records for patterns that are unique to their organizations.
2. "CA Technologies," *Wikipedia*, http://en.wikipedia.org/wiki/CA_Technologies. Neither the corporate Web site nor the *Wikipedia* entry for CA provides a comprehensive early history. Both lists are missing key early products, such as Dynam-T

(Tape Library Management), CA Sort (Sorting), and Dynam-D (Disk Space Management), but the text confirms that CA grew by acquisition and not by development. The result was a series of unrelated products linked only by marketing brochures.

3. For additional information on the issues surrounding component reliability and patching, see the Web site for the IEEE Reliability Society at: http://rs.ieee.org/.

4. See definitions at "Patch (Computing)," *Wikipedia*, http://en.wikipedia.org/wiki/Software_update; and "Upgrade," *Wikipedia*, http://en.wikipedia.org/wiki/Software_upgrade.

7

Postwarranty Hardware Maintenance

INTRODUCTION

This chapter deals with the options available to the buyer for postwarranty maintenance. Options explored include both the advantages and disadvantages of the various approaches. Readers will also learn about the many misconceptions about warranties, which should lead to a better understanding of how to buy appropriate support.

LIMITATIONS OF WARRANTY

The first line of support for most product problems is the original equipment vendor (OEM) or the author of the software. The OEM/author is expected to be the best resource for problem resolution because they built the product and presumably have more detail and experience with the product than any other entity.

The most common repair and support option for hardware offered to buyers is the OEM warranty. The term of the initial warranty varies by product and market. Because most manufacturers want to show as much revenue as possible in each fiscal year, most true warranties are structured so that the greatest possible amount of revenue can be posted. Unknown costs for delivery of warranties are projected and held aside in a "Reserve," the details of which are supposed to be clear in the reporting of public companies. Pressure from investors to show as much new revenue as possible creates an incentive to reduce the amount of money held in reserve to be as little as possible. There are two ways to do this: one is to lie about the potential costs of warranty exposure and the other is to book as short a warranty as possible. As a result, most *true* warranties are 12 months or less, and everything else is added as a prepaid service agreement even if the end user is told that the warranty is 3 years. Almost all buyers need to consider product life and support beyond 12 months, so how warranties and other support options are delivered have a wide applicability.

Not all warranties are the same or even cover the same types of problems. The attitude of the OEM toward extensions of warranty tends to break down into two categories: (1) the OEM that wants to make money by providing repair services and (2) the OEM that is not interested in repair as a business activity. It is obvious that those OEMs with an interest in generating revenue from maintenance contracts will arrange their policies to their advantage. Those with no vested interest in selling maintenance and support contracts will organize themselves differently and readily provide access to service parts, diagnostic routines, and updated microcode and firmware (also known as defect support) without restriction.

It is also a fact that not all OEMs are actually the designers and manufacturers of the product. Many vendors have a product built for them to market to their specifications. The major differences between such products are the physical presentation of the case and possibly some nuances of the interfaces.[1]

The support paradigm for such products is confusing. Many products brought to market are nearly identical internally to those marketed by others, but such information is rarely divulged. The actual need for break–fix service is essentially identical but closely guarded by the marketing entity.

For the end user, it may be useful to research the entity of actual manufacturing to better negotiate the terms of any support and break–fix (maintenance agreement). If an end user knows that the componentry of products X, Y, and Z are identical, then the options for repair and maintenance agreements are wider by a factor of 3 because parts can be purchased under a variety of labels and the technicians skilled in one brand of repair will almost certainly be able to repair another.

Voiding Warranty

Most warranties include limitations on support (hardware or software) based on the operating conditions of the equipment, such as equipment not being operated within specified tolerances. For example, many cell phones have moisture indicators that are used to prevent users from making warranty repair claims for units that have been dropped in the toilet (a common occurrence). There are limitations on many other products for other conditions such as exposure to excessive heat, which is known to be annoying to users in hot climates. Some clever OEMs have taken to boasting of their products' wider set of tolerances as a competitive advantage.[2]

Vendors also disclaim warranties for unauthorized modifications or sometimes the use of unallowable attachments. Nor will vendors support equipment that has been physically abused outside of "wear and tear," which is itself a difficult term to use in litigation. In cases where machines have been damaged by fire, flood, wind, and so forth, the expectation is the property insurance for the location in question is adequate for the replacement of products that cannot be easily returned to service. Most larger organizations have specialized insurance for data center equipment for this reason, which is tied to the cost of replacing equipment and not necessarily buying new equipment.

There are opportunities to negotiate terms of warranty, particularly when it comes to allowable and supportable attachments. Manufacturers may prefer not to permit the attachment of parts from other vendors, but when faced with having a purchase order, or not, they can capitulate particularly when the support relationship for the outside part is clear. The real issue of concern to the vendor is that they will not be required to diagnose and repair or patch problems that are not of their making.

Modification of contracts intended to allow "unauthorized" modifications to hardware are more negotiable than for software. Using automobiles as an example, one can usually make cosmetic changes without voiding the warranty but cannot ask the manufacturer to support aftermarket parts. The software equivalent is to "tweak" the timing on the engine control module (itself a chip), which is almost certainly a violation of the warranty. Yet playing with the programming on a chip should not void the warranty on the transmission.

Purpose of a Warranty

The real purpose of a product warranty is to support sales and marketing. The longer the supposed "warranty," the more importance it has as a marketing tool. For example, the Honda auto brand has stayed with a

36,000 mile warranty when competitors such as Hyundai and Kia are offering 100,000 mile warranties. Honda has the reputation of not needing service at all so Hyundai and Kia are using very long warranties as a marketing tool to overcome reliability objections to their products. *If a product were not prone to failure, then no warranty would be needed and none demanded.*

Using extended warranties as a marketing tool is not without costs. The seller of the vehicle (or other item) must make accounting adjustments in revenue reporting to cover the probable costs of service delivery during the warranty period. These allowances should be clearly shown in the annual reports of the manufacturer in the fine print. Warranty claim rates and accrual percentages should be publically available for all and even for many private companies either in their annual reports or in tracking publications such as *Warranty Week* (www.warrantyweek.com).

Investors are interested in the accuracy of percentages set aside for warranty accruals as they do not want to see adjustments to earnings based on inadequate or overly generous warranty allowances. It is clearly disadvantageous to the investor to keep the manufacturer from booking the revenue of the sale for years to come yet the marketing entity needs to offer the extended contract. The technique used by many in the technology space to sell the appearance of a long-term warranty without adding more accruals and adjustments is to internally transfer the revenue and the risk to a separate profit center, subsidiary, or division that manages all warranty service. This function can even be outsourced to a separate company. Members of the Service Contract Industry Council (SCIC) are major players in the provision of warranty services on behalf of manufacturers.[3]

There is significant profit margin built into these longer-term warranties, and they are not free to the customer although the vendor may position them as such. Multiyear warranties are not free, because they cannot be free. There is a direct cost to the manufacturer to make guarantees of any kind, including parts availability, so the costs must be reported somewhere in a public company. Vendors that are private companies can make whatever representations they like, but they still have a cost. In cases where vendors might be on shaky financial ground, the multiyear warranty is only as valuable as the vendor is capable of surviving to fulfill it.

Buyers would do well to demand a breakdown of the real warranty period and the embedded price for the multiyear service contract that is presented as the multiyear warranty. The best time to make this request is during the request for proposal (RFP) process and not as an afterthought.

Sales representatives are unlikely to be prepared for this request and the first reaction will be to balk at providing such detail. Buyers should expect to need to press for this information, which does exist as it is the way that vendors build their pricing. Consider the following lines of reasoning:

- Warranty accruals made by public companies are public. Request the accrual percentage.
- Postwarranty service contract pricing is often expressed in a percentage of original equipment cost (OEC). Request the percentage.
- Accounting for purchased equipment must separate services from hardware. Prepaid service contracts are not part of the depreciable value of the equipment. Request the breakdown.

For parts-only warranties, the direct costs to the OEM can be quite small, so some vendors may legitimately offer a long warranty on parts without having to make a large declaration of costs in their financial reporting. For most in the parts replacement business, the largest costs are logistical and related to shipping and processing. Users should be aware that the costs to ship a part are born by both the warranty provider and the user. Even if the freight is covered, users must spend their time to identify the problem, make the service claim, prepare the item for shipping, and then restore the unit to service by reinstalling the failed part.

Warranties involving labor are very costly. Warranty services offered with a labor component of longer than 1 year are certainly a packaged offering. These agreements are intentionally marketed and packaged to keep buyers from negotiating the uplifted warranty elements. Revenue from these uplifted services goes to the services side of the OEM business, which is usually a high margin profit center. The margins for the services side of publically held manufacturers are boasted to investors in public filings.

Widely distributed products, such as laptops, cell phones, tablets, and printers, are often sold with a parts warranty standard and an optional on-site warranty at a substantial additional cost. The most common standard warranty for distributed products is a return to depot repair offering, which requires the owner to ship the product to the designated repair location and await its return. As this is inconvenient and lacking in instant gratification, many big box retailers have created their own prepaid service offerings, such as "Geek Squad" or additional warranties sold at the retail point of purchase, which are in fact underwritten by specialist insurance companies, not the OEM.

Carry-In Warranty

Carry-in service centers have been common in the consumer electronics industry for decades. If the service center is not conveniently located, one is asked to ship the broken product to the closest service center where the repair is made in, or out, of warranty. Mobile products are increasingly serviced by the original retail outlet most often by swapping the unit and transferring the data. This dramatically improves the turnaround time for carry-in service from days to minutes and is clearly preferred by consumers. Some OEM-controlled retailers add to their profit margins by selling uplifted service plans, training, and accessories, as well as being a showroom for their latest products intended to replace models just a year or two old.

Major consumer products retailers have purchased service entities to add to their suite of offerings, for example, Best Buy owns Geek Squad. Others, such as the late CompUSA, offered retail services on a fee basis at their stores. The entire direct retail service model is undergoing change along with the challenges of retail operations in general.

Warranty Pricing

Warranty terms are not a measure of equipment quality, reliability, or durability. As discussed earlier, warranties are marketing tools.

Warranty pricing to the buyer is unrelated to the need for repair of the equipment. Buying an extended warranty buffers the buyer from buying parts, but little else. The real delimiter of quality need for service or repair is the mean time between failure (MTBF) of the equipment, almost always unknown by the buyer.

Consider that for the general quality of equipment in common use, OEMs are setting aside 3% to 5% of their entire sales revenue to cover their wholesale costs to service brand new equipment. They would not set aside these sums if they did not expect to spend it. Since buyers are paying for service at retail, and not wholesale as set forth in the accruals, at the very least it means that buyers of electronic technology should set aside double the wholesale costs as a rosy minimum for service each year. The accrual reconciliation process for OEMs covers only the initial 3- to 12-month warranty period, so these figures are undoubtedly the best case scenario.

Buyers of information technology (IT) products are so accustomed to failure that they routinely buy extended service contracts for technician

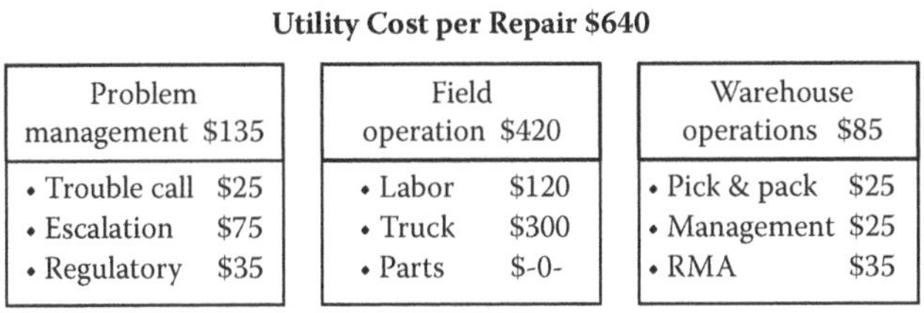

Utility Cost per Repair $640

Problem management $135	Field operation $420	Warehouse operations $85
• Trouble call $25 • Escalation $75 • Regulatory $35	• Labor $120 • Truck $300 • Parts $-0-	• Pick & pack $25 • Management $25 • RMA $35

FIGURE 7.1
Utility cost per repair.

service at annual costs commonly 30% or more of the asset cost. On-site hardware maintenance is big business in IT.

In contrast, many warranty programs for distributed assets rely upon the owner to dispatch an employee to retrieve, pack, and ship the item for warranty replacement. In a corporate campus this is relatively simple, but very problematic for distributed products. Some electronics, as with electric utilities, are deployed in settings that have dramatic safety issues. Buyers of products that have exterior and distributed support issues need to consider the additional expenses of dispatching trucks and technicians to pull and replace equipment regardless of component value.

Figure 7.1 shows that dispatching a truck and crew to pull and replace any asset due to any type of part failure can be far more costly than the unit cost of the original item. Current deployments of smart meters and other technology applications are a case in point. Most meters are low-value products (less than $200), but the cost to touch them is clearly far larger. In these settings, in addition to the replacement of products that arrive DOA (dead on arrival), the warranty provided by the OEM is the tip of the repair iceberg. The real costs are on the utility side.

ORIGINAL EQUIPMENT MANUFACTURER POSTWARRANTY SUPPORT

Keeping in mind that the actual warranty period for configured equipment is 12 months or less, postwarranty support must be considered for the majority of the lifetime of every technology asset. The most common, and frequently highest cost option, is to enter into a postwarranty support agreement with the OEM.

OEM offerings are convenient but not comprehensive. By design, OEMs often limit their service options to high-density locations, deliberately avoid supporting old equipment, and frequently are inflexible with terms and conditions, particularly those that involve support of products in a multivendor setting. Since most IT organizations have a history of involvement with multiple vendors, the multivendor setting in which each vendor has its own contract for its own branded equipment is most common. It is unusual for a single vendor to be providing in-warranty and postwarranty support because almost all companies will have a wide variety of equipment deployed representing different eras of computing preference, supporting specialized applications, or a legacy of equipment stemming from acquisitions over time.

Escape from such complexity is attractive in the form of a "Four Walls" or other comprehensive multivendor support agreement. Whatever entity packages the combination, none are servicing the entire spectrum of equipment using their own resources. The packager provides the services for which it is most adept and manages subcontractors for everything else. Users of such services are paying for multiple levels of profit margin in order to have the convenience of "One-Stop Shopping."

OEMs typically offer the same type of services for their warranty and postwarranty options. A manufacturer that offers only a depot repair initial warranty is unlikely to offer a postwarranty service inclusive of labor. Conversely, OEMs with parts and labor warranties generally offer an extension of such services as a business and do not usually separate parts and labor within the agreement. Many OEMs further complicate the selection of postwarranty support with demands that hardware and software support agreements be combined as a condition of any service.

Revenue from the manufacturer controlled maintenance organizations with an on-site labor option is usually posted to a separate profit and loss statement (P&L) within the OEM business. The profit margins on services are occasionally so much higher than for the equipment sale that equipment sales could be considered a "loss leader" to capture the service contract.

OEM organizations that offer on-site hardware maintenance service are in the business to make money. They have a profit motive to sell services. It is to their advantage to offer service options that bind service delivery exclusively to themselves. The marketing of break–fix typically includes some disinformation as to the difficulty of providing service as well as the risks of deploying other forms of service options, particularly competitive options.

Disinformation on the topic of hardware maintenance is generally directed at preventing buyers from seriously considering non-OEM offerings, particularly those of Independent Service Organizations (ISOs), also known as Third Party Maintainers (TPMs). The most common of these tactics are the concepts that:

- Only the OEM has access to service parts.
- Only the OEM has adequately trained technicians.
- Only the OEM understands the complex interplay between hardware and software.
- Only the OEM has analytical skills.

This is only partially true. OEMs control access to service parts but can decide how quickly they wish to respond to end-user requests for parts. End users can demand access to service parts on a timely and reasonable basis. Further, once a product is more than a year old, OEMs no longer completely control their own parts market.

OEMs have some direct employees as trained technicians, but most, if not all OEMs, frequently subcontract to qualified technicians as flexible labor. It is costly for any business to keep employees on staff waiting for service calls that cannot be predicted. It takes a very large OEM with a very concentrated customer base to be able to keep technicians busy in any given locale. The skill level for IT technicians is also less complex than in decades prior. Most products are designed to be easy to repair by pulling and replacing parts. Using remote diagnostics allows the service provider (not exclusively the OEM) to dispatch the technician with all the known parts already in hand. It is only when the parts are replaced and the unit is not restored to service that there is any practical advantage for the OEM to be in control.

Hardware and software are complex, but the design functions are mostly separate and the responsibility for corrections to design defects is also separate. As shown in Figure 7.2, the OEM creates the specifications for a product it wishes to produce. The specifications usually include the choice of operating system, unless the product requires a custom operating system (OS). The selection of the chipset is also determined at this time. Even if the chipset is under development as a new product, the remainder of the machine must still be designed so that the chipset can be included at the time of assembly. When problems are identified either during or after assembly, the required corrections are made by the responsible

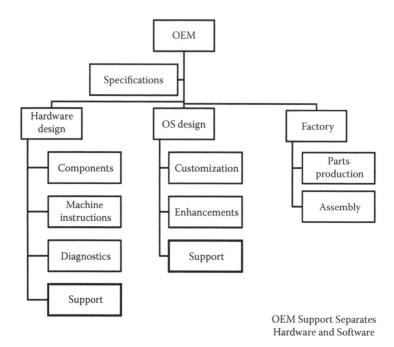

OEM Support Separates
Hardware and Software

FIGURE 7.2
Support role diagram.

group, which is almost always either the hardware designer or the software designer, but rarely both.

The reason that there is a solid differentiation in support roles is that for commercial computing, the hardware designer, often a subcontractor, designs the printed circuit board (PCB) that will be used for assembly along with any machine instructions that are needed to operate the machine. The designer is responsible for the interaction of components in the design and is therefore responsible for any errors in interaction. If design problems are identified during or after assembly, the hardware designer is the only party capable of making corrections. Unless the machine operates as specified, any subsequent layers of software, such as an operating system or application, will not operate correctly, if at all.

Problems of assembly, such as poorly soldered connections, are not the responsibility of the hardware designer, unless the board design itself causes problems with assembly.

Software engineers, while perhaps involved in setting specifications for machine instructions, have the task of taking a machine that already passes current correctly and making the machine "do something." If there is no software, including embedded software, digital machines do not magically perform tasks. Simple and repetitive tasks, such as included in a home thermostat, are almost always using embedded software. The designer of the thermostat software is responsible for any errors in the code. Because the integration of hardware machine instructions and software is tight in simple machines, there may be overlap in design skills. However, the more complex the machine and the more layered the software, the more differentiated the roles. As a result, in most commercial multifunctional IT products, the software designer is quite separate from the hardware designer.

Even for those OEMs that fully control the design process, the support teams that are responsible for actual problem determination and corrections to hardware defects and machine instructions at the design level are completely separate from the software support teams that diagnose and correct design problems with the OS or other software layers. Support is further differentiated by employment and management structures with the OEM organization, which confirms that the skills are viewed as separate by the OEM as well.

INDEPENDENT SERVICE ORGANIZATIONS

Independent service organizations (ISOs) are not just direct competitors for OEM break–fix service, they are the glue that hold both postwarranty and multivendor service agreements together for the OEM. OEMs make constant use of ISOs as subcontractors for locations and equipment that they deem unprofitable to service directly. Among the ISOs are specialist independent companies that repair and restock equipment for warranty support. There is a large ecosystem of independent companies that make the OEM to end-user support agreement work. This also means that there is an opportunity for all end users to investigate opportunities to take service contracts directly to the service provider instead of using the OEM as an intermediary.

Once an OEM declares "End of Life" for products, it no longer offers a break–fix service program and the only resource for support is the ISO. OEMs without a labor force of their own routinely subcontract to ISOs. In an ironic twist of marketing FUD (fear, uncertainty, and doubt),

OEMs offering multivendor support contracts largely prefer to deal with an ISO to support a competitor's products than to subcontract directly to the competitor, since they expect the competitor will do more damage in the account than good.

DOWNSIDE OF INDEPENDENT SERVICE ORGANIZATION OFFERINGS

The most obvious risk is that the selected ISO does not perform. This is always a consideration in selecting any product or provider, and most competent purchasing organizations are adept at checking references and other cross-checks for service quality. Unless the OEM refuses to accept a return to service, it is always possible to humbly return to the OEM service. This is no longer as common as it once was as OEMs have discovered that they can tell current users they cannot return to service without a penalty. This is not always the threat that it appears, as within the OEM organization there are competing interests for the profits generated by the service organization.

- The service organization within an OEM is a profit center. They will often take service contracts, which are in conflict with some objectives of the new equipment salesperson who is not compensated on the services sale unless it is part of the equipment purchase. End users have tremendous negotiating power when considering a return to ISO service.
- If an ISO is not performing, the most likely resolution of the conflict is to change to a different ISO, not return to the OEM. The reasons for using ISO support and not OEM support are unlikely to have changed. The OEM most likely does not offer service for this category of equipment or has already burnt its relationship with the end user.

The second most obvious problem comes down to novelty. If products are less than 2 years old, there may be some difficulty for ISOs in acquiring parts and skills. As most new equipment is supported under at least a 1-year parts and labor warranty, the ISO typically has its first experience with the newest models in the second year of product availability. If availability of machine code patches and fixes is limited by the OEM, users would do well to buy a second year of OEM support to assure access to the vast majority of applicable machine code patches and upgrades before moving off OEM support. By the end of the second year, most products are as stable as they will be. Further, many products will have already ceased production and while support is still offered, the volume of novel problems will be marginal.

As with any relationship, there is some risk that a ISO (or OEM) salesperson will overpromise and underdeliver. This is a normal fear and for this reason most ISO contracts begin with a small subset of equipment. Only after the ISO has proven its capabilities is the suite of support widened. Despite all of the aforementioned advantages to the OEM, the ISO support model has proven itself to be credible competition throughout the IT industry.

PARTS WARRANTIES

OEMs have difficulty offering lengthy warranties for parts because they eventually run out of parts. The velocity of new products and new manufacturing techniques dictates that 3 years is outside the OEM's ability to comfortably maintain spare parts. Beyond 3 years, even with scavenging the used market or having parts repaired at the circuit board level, providing parts is an inevitable challenge. Users seeking large volumes of spare parts can order spares with their initial orders for their own use.

Nor do OEMs want to extend the comfort period for users by extending equipment warranties, as the fear of breakdown is an important driver of new product sales. It suits OEMs to sell replacement products to drive new revenue. Lack of parts fits nicely with the control that the OEM wishes to have over the replacement cycle.

Most electronics designed today are assembled to order in Asia in large production runs sometimes in the millions of units. Each run is shipped in its entirety through distribution and includes all the replacement parts ordered for that run. By the time the equipment is actually delivered, the production line may have long ago moved on to another product, probably for a different vendor in a different industry. Running batches of parts for replacement use is rarely done by the OEM. Sources for large volumes of replacement parts exist, including custom runs, but these are not usually OEM services.

MACHINE CODE AND EMBEDDED SOFTWARE

Access to updates to embedded software or machine code in postwarranty service contracts is highly variable. Updates are needed in case there were any corrections made to the code after the machine was shipped (or after the expiration of the initial warranty). Access to original code is also needed in the event that the code needs to be restored. There are several types of functions, such as diagnostic code and service access codes (passwords), which are lumped into this category and must also be available for proper repair.

In situations where the OEM blocks access to machine code to parties other than the OEM, users are not able to avail themselves of any postwarranty options outside the OEM. Such blocks extend to users being unable to access the library of machine code to restore code that had already been provided on the machine at the point of purchase. Treatment of access to the entire library of patches and fixes to the machine is the lynchpin for access to postwarranty service options other than the OEM.

Limits on machine code access are a relatively new tactic in wide use only after the Digital Millennium Copyright Act of 1998 (DMCA) was passed. At the time the DMCA was written, most fixes to machine logic errors were delivered using physical engineering changes (ECs) or by distributing patches and fixes on media (such as tape or disk) shipped through the mail. If there were charges, the costs were for postage and nominally for the media.

The closest the DMCA comes to addressing machine code is to affirm the rights of the user to make a backup copy of licensed software without being in violation of copyright law for making a single backup copy. There are no references in the DMCA to machine code or any acronyms. It is clear that the DMCA did not at all anticipate the existence of machine code as anything separate from the hardware.

Policies that thwart access to all forms of machine code are currently at issue in several legal and legislative settings. Until such time as the DMCA is amended to address these issues, or some of the litigation already in the courts provides precedent or state laws are amended to clarify how consumers and businesses can access such code, users are on their own to negotiate how machine code is to be treated directly with the vendor. It is essential to make these changes before signing any contracts.

Digging deeply into the policy of the OEM often reveals inconsistencies between how updates are actually delivered and marketing claims.

When faced with policy proclamations from the sales force, it should not be assumed that the sales force is correct, so the actual policy document as issued by the home office should be reviewed. There are many cases where an overly ambitious sales team has overstated or misstated the policy goals. Large corporations with marquee names are not immune to this and may possibly be more prone to inconsistency than smaller and more tightly held OEMs.[4]

In situations where the marketing claims are untenable or questionable, the marketing team should be pressed to provide clarification on all points signed by an executive on letterhead and ideally with proof of the credentials of the executive to bind the corporation. This request almost always flushes out the hype. Users should be prepared to put a time limit on the official requested response since such a letter is either easy to prepare (in which case the policy is correctly stated) or the policy cannot be confirmed (in which case the letter will never be written). Lack of response and official silence is confirmation that the OEM is not confident that it is within its rights to press the policy and additional pressure may resolve the issue in favor of the user. Corporate counsel should review any of these situations.

SELF-REPAIR

A wide variety of equipment can be easily repaired by minimally trained technicians or even end users. Many products are designed so that low-level skills are sufficient to pull and replace common components. Taking advantage of these opportunities for self-service involves having a suitable labor source, access to parts, and a system for managing the logistics of return merchandise authorizations (RMAs) and restocking. Many organizations combine self-service with ISO services and board-level repair specialists to manage elements of the self-service process without putting too much burden on the end user/employee.

The largest problem confronting end users when evaluating a self-service option is the relationship to the OEM warranty. Some OEMs, particularly of highly commoditized products such as PCs, will authorize end users to perform self-repair within their warranty program. Typically, the authorization for internal break–fix is limited to the enterprise and cannot be offered as a for-profit service. The self-service technicians must meet certain training standards and the organization must provide all the management and logistics for the OEM. There are often volume requirements as well; the buyer of the equipment must be a regular and sizeable user to be eligible to create an in-warranty repair program.

Another variation of the self-service options is outside the OEM warranty. The end user develops the skills to pull and replace equipment and acquires a stock of spares. As the failed units are repaired (sometimes by the aforementioned board-level specialists), the stock of spares is continually refreshed. Additional spares enter the supply chain as older units are migrated to newer models. This method is common with equipment that sold without a labor warranty where there is high volume of identical assets that can be easily swapped.

Not all OEMs support self-repair even for high-volume, low-unit cost equipment. Apple, as an example, only allows end users with thousands of units (it was 200 and has since changed) to be trained and "authorized" for self-repair. Many others will void all warranties if the case is opened or "cracked" by any party other than themselves.

Self-service is often overtly thwarted by OEMs through limitations of parts purchases, an annoyance, or more important, denied access to machine code for patches or reload. Many common items are shipped with security features intended to prevent theft of information in the event of a unit theft but have the effect of also blocking the equipment owner from restoring equipment without a service contract from the OEM.

Most products that handle credit cards have such security features, which are an increasing repair frustration as credit card use proliferates into very small transaction sizes including vending machines, self-service kiosks, and all traditional retail. OEMs in these products have resisted any support of self-service or ISO service under the guise of protecting "security," but these security issues belong to the merchant. Very large merchants have successfully taken control of their own break–fix for these devices.

Repair of consumer class products is increasingly popular with owners comfortable with opening cases and replacing simple parts. Self-repair by individuals is aided by Web sites that provide parts, how-to guides (often as YouTube videos), and nonproprietary tools. Such Web sites and businesses are constrained from providing copies of manuals (possibly a copyright violation), and are further limited in providing tools that allow access to cases that are not intended to be opened by the consumer. At least one business, iFixit (www.ifixit.com), avoids the potential risk of copying manuals by building its own how-to guides. This greatly limits the scope of products it can cover to only the most popular models of equipment.

The tool issue is also a copyright issue stemming from a different part of the DMCA covering what had been a VCR media piracy issue in the 1990s. The language, called "Anti-Circumvention," was written to prevent vendors from creating tools (software or hardware) to enable users to modify their VCRs to make copies of tapes and sell them.[5] Unfortunately for the rest of the technology industry, the wording is not limited to VCRs and could potentially apply to any tool (hardware *or* software) made to access any technology product. The same anti-circumvention provision of the DMCA issue is at the heart of the copyright restrictions on cell-phone unlocking. The wireless industry has reluctantly agreed to allow unlocking under the threat of legislation by the U.S. Congress, but the larger issue of access to embedded code and restrictions on tools remains unaddressed.

AUTOMOTIVE REPAIR: CONVERGING ISSUES

The automotive and IT industries have been converging for some years with increasing use of digital technology in vehicles. Although most owners do not see the technology, digital parts have taken over many functions, from controlling engine time (engine control module), to controlling braking (ABS brakes are computer controlled), to self-diagnostic messaging (tire pressure, oil changes), and of course the dashboard display. Manufacturers have found that they can push buyers back into their dealerships for post-warranty service by limiting access to parts, tools, diagnostics, manuals, and microcode updates for digital parts. The more digital parts, the more control the manufacturer can exert.

Repair of automobiles is now guided by the output of diagnostic equipment attached to a port under the hood that links the external equipment to the sensors and monitoring equipment inside the vehicle. Repair technicians for other computerized devices use the same types of products. Once the malfunction is identified by the diagnostic tools, the technician can pull and replace the malfunctioning part and quickly restore the unit to service. The break–fix process for vehicles and computers is very much the same, minus the grease.

There have been successful efforts by the Automotive Aftermarket Industry Association (AAIA) to pass legislation guaranteeing consumers an automotive right to repair first, which was passed in Massachusetts in August 2012.[6] The auto industry worked out a Memorandum of Understanding

(MOU) with the aftermarket industry covering the same points in the Massachusetts law to be applied nationally.

The five major problems addressed in the MOU for automotive right to repair legislation include removing:

1. Restricted access to service parts outside the dealer network
2. Restricted access to "proprietary" tools
3. Restricted access to diagnostic tools (software and interface problems)
4. Restricted access to microcode and firmware updates for component parts
5. Restricted access to manuals and schematics

These problems are the same for repair of digital electronics in many other settings. Everyone seeking to control repair for their equipment needs to be assured access to all five major interlocking functions. Depending on the industry, some, or all, of these functions may be limited to the OEM repair staff, thus preventing independent or self-repair.

FIVE REQUIREMENTS FOR REPAIR

Repair of digital products, in or out of warranty, has five common requirements (Figure 7.3). Limitations on any of the five heavily limit what can be done outside of the OEM. The most central limitation is access to machine code and associated updates. None of the other limitations has the same impact and most have some workaround. Without machine code, no product operates and therefore has no value other than component parts.

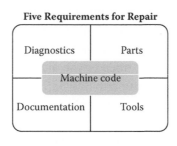

FIGURE 7.3
Five requirements for repair.

Diagnostic Tools, Diagnostic Software, and Error Codes

The first step in most repairs is to run diagnostic routines supplied by the OEM to identify the cause of the fault. Most equipment is provided with some self-checking routines, which are intended to display at least an error code so that a technician can work more efficiently. (Technician time is costly so the OEM has an incentive to be informative to minimize costs for warranty support.) Following the attempted repair, diagnostics are typically rerun to confirm the repair is complete. If the diagnostics indicate a problem still exists, the job is not complete.

Diagnostic software, either onboard or external, is not always provided as part of the equipment purchase. In the auto industry, the diagnostic software may be included in a separate device that must be purchased to read the diagnostic information from the vehicle. In this case, diagnostics are both tools and software. In cases where diagnostic services are provided by external equipment, the OEM may refuse to sell such equipment to equipment owners or independent repair organizations, thus pushing repair back to the OEM exclusively. If a developer decides to fill the void by manufacturing an independent set of diagnostics, the prohibition against circumvention in the DMCA will likely be used to litigate against the provider of the equipment. In most cases, the threat of litigation is enough to dissuade small companies (entrepreneurs) from entering this market.

In most IT equipment, diagnostic routines are provided with the machine and accessible using standard input/output devices. Although this eliminates the need (in most cases) for an additional purchase of diagnostic equipment and specialized attachments, the manufacturer can still block access to diagnostics by outsiders in several ways. First, the OEM may deliver diagnostic code updates only under an OEM maintenance agreement. Second, the OEM may deliver a set of diagnostics to independents that are different than those for their own technicians, narrowly meeting legal requirements in some jurisdictions. Similarly, some OEMs block access to corrections to the diagnostic code itself. Third, the OEM may lock access to diagnostic routines using software locks or quasi-hardware locks that are delivered exclusively to OEM technicians. Last, and extremely common, the OEM simply claims that the diagnostic software is their intellectual property (IP) and the implied threat of copyright litigation generally ends the attempt to access diagnostics without the permission of the OEM.

Tools

Repairs usually require some physical tools, such as a screwdriver, to open cases to access parts. Within the machine, there are commonly other tools needed to remove and replace failed components. Some machines are designed to be accessed by the consumer with common tools, and others are designed to be difficult to access without specialized tools that are deliberately uncommon to thwart repair by the owner or any non-OEM technician. Buyers should request that proprietary tools be provided as part of the purchase, in much the same way that small wrenches are provided by IKEA to facilitate installation of their products.

Documentation

Manuals are less commonly delivered with sufficient information to guide an owner toward repair. Most consumer products are shipped with just enough information (in many languages) to guide the owner to install or operate the product. Manuals with sufficient detail to support repair are not usually provided to the end user but are created for technicians dispatched or authorized by the OEM. Many forms of documentation are no longer delivered in hard copy and are instead posted on Web sites. This allows the manufacturer to tightly control access to documentation by placing documentation behind firewalls, behind user registration, and behind service entitlement databases.

In years past, documentation was provided in hard copy to the buyer and the documentation, including logs from field engineers, was expected to be passed from owner to owner or location to location. Although the documentation was itself copyrighted, the owners had a legitimate copy and could provide that copy to any repair organization of their choice. This is no longer the case and users must now make sure to negotiate for access to manuals in the procurement process. Users that cannot acquire service documentation are not able to select repair services on their own terms.

In addition to Web sites and other blocks, OEMs are reluctant to provide schematics, but schematics are needed for many types of repair. OEMs argue that they should not be forced to give up "trade secrets," which may or may not be present in some forms of documentation. Users should not be shy about requesting suitable documentation; the OEM can choose to create suitable documentation and still protect trade secrets. Otherwise, owners will give up their right to repair products outside the OEM offerings in exchange for capitulating to the OEM assertion that they might be inconvenienced by having to modify documentation to meet the needs of the buyer.

Manuals are only partially a copyright issue. It is illegal to make copies of intellectual property, but not all manuals are agreed to be sufficiently creative to be treated as IP. Unfortunately, since each manual for each product might have to be judged separately in the courts, the default position on manuals is not to copy any manuals. Original copies of manuals can be included in secondary market transactions, but many owners discard manuals and are often unable to provide them for use years past the original purchase. The current delivery model for manuals is for manufacturers to post them on Web sites, but there is no obligation to do so.

Many repair companies have developed their own internal versions of manuals to guide their technicians. These are no more available to owners than those of the OEM. Consumer products in wide use are frequently pulled apart and repaired by hobbyists, who in turn post their processes online. If any of these "authors" were to make their manuals for sale, they would own the copyright to the manual and can offer it on any terms they select.

Not every product in use has a manual available. Before purchasing a product, it is wise to check availability of all manuals and require repair manuals are part of the purchase. Copyrights last for 75 years, which is an impractical limit for technology products, so buyers need to make sure that they protect the value of their investments by keeping all legal versions of manuals until the product is resold or recycled, or copyright law is reformed.

Service Parts

Service parts are necessary to complete a repair in a timely fashion. Without a fully functional replacement part at hand, many repairs could take hours of highly skilled technician time, if the repair can be made at all. OEMs have a vested interest in making repairs both fast and cheap, so most products are built to use plug-and-play spares. Stocking and availability of service parts from the OEM is completely controlled by the OEM. Not all OEMs are cooperative when it comes to selling service parts to entities they view as competitors, such as independent repair companies as well as end users.

Customers, and particularly large buyers, have the power of the purchase order to put requirements for access to service parts into their agreements.

There are two basic problems with service parts that can be addressed by customers: price and timely availability. Service parts are rarely priced at cost. The newer the part and the more limited the availability of other options (such as used parts), the more likely the part will be highly priced. Users have been known to require parts pricing limits based not on an arbitrary "List" price, but based on the prices commonly charged, net of all discounts and rebates, to authorized service providers. The exact terminology needs to be provided by counsel, but the concept about the net price is to keep the OEM from setting one set of fake prices but deeply discounting the part to favored parties.

Timely availability of service parts is a challenging issue to negotiate. OEMs can pretend to be out of stock on parts whenever it suits them. The OEM does not want to make it easy for a third-party competitor to have ready access to parts, so the delay tactic is extremely common. Buyers need to anticipate that OEMs will be uncooperative with respect to filling orders for service parts that come through their designated repair agent, and may be better served by ordering the part directly. Even with the direct order, the OEM may still be reluctant to ship parts, so buyers need to put some teeth in their contracts requiring parts availability in some way, such as on an even footing with the OEM authorized repair channel (or the OEM directly).

Aftermarket providers are a healthy check and balance on OEM intransigence on service parts. The technology parts market is multifaceted including parts scavenged from used (or stored) equipment, parts purchased from the original commodity parts manufacturer (often in Asia), and in some cases custom manufactured parts. The velocity of technology parts changes is far greater than that of air filters or lug nuts, so there are not often third-party manufactured parts or private label parts to be used in lieu of original parts. In the consumer markets, non-OEM parts are far more widely available.

Because of the alternatives available in the supply chain for IT parts, service parts availability is the least important of the five requirements to postwarranty technology repair because of the many options to workaround OEM restrictions.

Machine Code

It is not currently common for equipment owners to have the ability to backup machine code on a regular basis. Such backup and restore rights are included in the current U.S. Copyright Code,[7] but are not practical so the equipment owner is not clearly in control of machine code. This makes it difficult for an equipment owner to repair their own equipment or hire someone to repair the equipment, unless the OEM has provided a complete library of patches and fixes on its Web site for use in the event of a problem.

Blocked access to the library of machine code, regardless of intent, is the single largest problem facing independent repair companies from providing competitive service in any industry. Machine code controls the internal function of the hardware regardless of the applications installed. Machine code is not always perfect, so it is occasionally patched and fixed after the product is produced. These fixes impact the hardware operations of the product and must be applied or the product will not work as intended.

Furthermore, machine code also controls access to features at the machine level. If the original product was shipped before new hardware features were added, such features have to be included in the instruction set so they can be accessed. (This is different than device drivers, which can be added to the operating system.) For example, a line of servers designed in 2005 would have had machine code appropriate to the chipset speed and cache. As newer models were developed, faster chips and larger capacity cache would have become available. The older machine code would not have been set to operate at the higher speeds, and probably did not include the ability to address the larger memory and cache sizes. Machine code updates would be needed to match any chipset upgrades that the user might want to make. Blocking access to these types of functions is self-destructive for the OEM, unless the OEM intends to block attachment of used products.

If the OEM does not provide these updates to all owners without regard to warranty status, then owners must purchase those updates from the OEM to be sure that their machine works correctly. OEMs have learned in recent years (since the DMCA was approved) that limiting access to patches and fixes to machine code to their exclusive repair services is their single most effective weapon against having to compete for postwarranty repair services.

An extensive discussion of machine code is found in Chapter 8.

SUMMARY

Warranties are the beginning, not the end, of the selection process for hardware maintenance over time. Options for postwarranty support are highly variable, and OEMs may make unpleasant requirements in order to continue to use their equipment or services. Buyers that take the time to control their access to the five major requirements for repair, shown next, will benefit.

1. Access to service manuals and documentation
2. Access to service parts on a timely and reasonably priced basis
3. Access to diagnostic code, both internal and external
4. Access to the library of machine code updates (patches and fixes)
5. Access to specialty tools, if needed

NOTES

1. Manufacturers in China, such as Viasystems (http://www.viasystems.com/), are the designer and manufacturer of products marketed by multiple vendors in the United States and elsewhere. Viasystems provides the circuit board design as well as the manufacturing and product assembly. Viasystems is only one of many firms in the industry.
2. See YouTube ("GALAXY S4 Factory Testing Reliability," http://www.youtube.com/watch?v=N7NVH_9o9BA#!; in Japanese) for the Samsung Galaxy X4 exploiting the poor operating tolerances of competition (most likely Apple) in situations of modest heat.
3. For a partial list of companies providing these services, see: http://go-scic.com/insidepages/membership.cfm.
4. IBM is currently in the process of implementing a new policy regarding machine code. The policy has yet to be clearly or consistently articulated or applied. HP has recently announced similar changes with even less clarity.
5. For the entire section of Section 1201 of the U.S. Code, see: http://www.copyright.gov/title17/92chap12.html#1201.
6. An article describing the legislative history of the Massachusetts Right to Repair Act can be found at "Motor Vehicle Owners' Right to Repair Act," *Wikipedia*, http://en.wikipedia.org/wiki/Motor_Vehicle_Owners'_Right_to_Repair_Act.
7. See DMCA Section 117 for details. "17 USC § 117—Limitations on Exclusive Rights: Computer Programs," Legal Information Institute, http://www.law.cornell.edu/uscode/text/17/117.

Section III

Technology Product Details

8

Machine Code and Embedded Software

INTRODUCTION

Machine code and embedded software are common synonyms for what has become the dividing line between tangible hardware and intangible software. This is not widely understood or appreciated, likely because as long as there are no problems with this very basic level of programming, users are blissfully ignorant. However, when contracting for hardware or software, ignoring machine code is an enormous error. Most of the manner in which machines are open or closed to repair is based on perception, sometimes false, about the nature of machine code and how it should be treated.

DEFINING MACHINE CODE

Machine code is a catchall phrase used by many original equipment manufacturers (OEMs) to reference the wide variety of routines that are not specifically or separately licensed but remain integral to the operation of digital machines. There are other descriptions including embedded software, embedded code, microcode, firmware, basic internal operating system (BIOS), unified extensible firmware interface (UEFI), internal operating system (IOS), programmable logic code (PLC), and a host of variants. For simplicity and general use, this book makes consistent reference to machine code, which is a common term in enterprise information technology (IT), and occasionally embedded software, which is more widely used for consumer class IT. All references to machine code are intended as a generic reference inclusive of all variants.

MACHINE CODE DIVIDES HARDWARE FROM SOFTWARE

One can argue if machine code should be considered "Intellectual Property (IP)" or hardware. Regardless of the final decision, machine code is the dividing line between tangible assets and intangible IP. There is no reason to believe that any firm legal delineation is going to be forthcoming, so users will remain in charge of which side of the argument they wish to take.

It is clearly a problem for buyers to treat machine code as intellectual property without either a price or a license agreement. Buyers should take advantage of the lack of clarity in proposals involving vague references to machine code to require reasonable descriptions, clear terms and conditions, and an associated value. Currently, there are no vendors providing any clarity on this subject, which buyers should take as confirmation that this is a wide open opportunity for negotiation. (See Figure 8.1.)

There are vendors, such as IBM, that have begun creating "licenses" for machine code, but these agreements are not yet clear enough for buyers to accept without investigation and negotiation. First, such licenses do not define the functions of the machine code being licensed, leaving the OEM the opportunity to make any claims at any time for any purpose. Second, there is no price associated with the licenses, leaving owners unclear as to how to account for such licenses in financial reporting. Further,

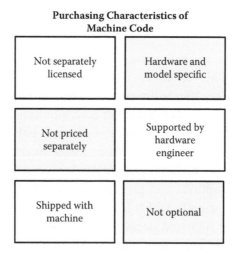

Purchasing Characteristics of
Machine Code

Not separately licensed	Hardware and model specific
Not priced separately	Supported by hardware engineer
Shipped with machine	Not optional

FIGURE 8.1
Characteristics of machine code.

maintenance pricing for other IP is generally tied to the original acquisition price of the licenses, but this is not clearly the case for machine code. Buyers need clarity on all these points to fully understand their obligations to the OEM.

Absent discrete and sensible licenses, there are other problems with how machine code is delivered. Machine code is both model and feature specific. Users of one model of machine cannot *pirate* the machine code of another and expect it to work. Each model of machine, or model series of extremely similar machines, is designed with specific machine instructions unique to the design and the available features that might be active. If a new feature is created, such as a new model of denser memory than had ever been produced, the machine code must be updated to *see* the additional memory addresses. Without the update, the original machine still works but the user cannot take advantage of higher memory capabilities.

OEMs use the machine-specific nature of machine code to discriminate against competitors as well. For example, when EMC came out with its early Disk Array products, IBM chose not to create a "native mode" attachment in machine code. IBM had a competitive storage product and despite the clear success of EMC in the market, IBM did not update its machine code. EMC had to design its products to appear to the IBM system as an existing IBM device. OEMs can, and do, refuse to provide machine code updates to users they wish to punish, such as buyers of non-OEM products, used OEM products, or users of independent support service.

If there are flaws in machine code, the impact is profound and impacts every machine built of that particular model. Other models are not affected, since their code is unique to that model. The hardware engineer (hardware designer) has to make corrections to machine code for the simple reason that the hardware engineer created the instructions as part of the machine design. Patches and fixes to machine code problems cannot be made at the factory during assembly nor by the designer of the operating system or developer of any apps. Machine code, at least for purposes of corrections, is treated as hardware.

Finally, machine code is shipped with the machine already installed. It is not possible for the end user to separately load machine code postpurchase since the machine itself includes the code needed to start up. Start-up code is deeply embedded in the hardware and if the machine does not start up, the machine is as useful as a brick. There are elements of machine code used in some security sensitive uses to self-validate condition at start-up and if the validation is unsatisfactory, the machine "bricks" itself.

The day may come when users are asked to license, and pay for, machine code routines for specific purposes. For example, machine code controls which cell phone carrier is used. Unlocking a cell phone requires access to the specific routines for that purpose. Until users have to license this code, it is entirely unclear who really has the right to modify the code. The decision on which parts of a cell phone are licensed and which are owned are being argued by the U.S. Congress as well as the court of public opinion.

The same core questions of ownership apply to other products and other functions described later. Regardless of how such code is treated—as IP or not—and if the manufacturer/author wishes to license the code, the user has to be party to the license. Most of the theme of this book is about disclosure and when it comes to disclosure, machine code needs a great deal more attention.

Purchasers are commonly unaware that machine code even exists. It is not discussed in sales presentations, not mentioned in brochures, and rarely discussed outside the realm of postwarranty support providers. General ignorance is not bliss; the entire industry is shifting support models to leverage the unique value of machine code into postwarranty service monopolies. So far, OEMs have resisted posting prices for machine code on initial purchase contracts, likely in order to avoid having to explain payment for things that used to be included.

Machines always include machine code or embedded software, as they would not operate otherwise. Machine instructions are the first instructions that activate when the "On" button is touched. Machine instructions start spinning the disk drive, seek the location of the operating system to launch the operating system (OS), and remain active to manage all the bit-to-bit instructions of making a computer compute. Even when simple machines are shipped without a separate operating system, the machine includes machine code. This is the core of the argument that machine code should be treated as hardware and "owned" by the buyer.

Because each new product has unique capabilities and design, each product has its own machine code. While there are undoubtedly subroutines that are repeatedly reused by hardware designers, machine code is rarely transferrable between product models. Developers of machine code do not fear piracy in the sense that machine code could be copied and proliferated. The only buyer of copies of machine code would be the buyer of the machine, the same party that already has the code.

OEM concerns about violation of IP in the world of machine code are fears about user modifications to machine code intended to enable functions not intended by the developer. In a case of "jailbreaking," Apple argued it needed to "Control the End User Experience," and the U.S. Copyright Office ruled, wisely, that buyers of the product get to decide how they use their purchases. Behind the claims of end user experience lies the real concern: that the user can modify machine code to enable functions the OEM wishes to provide exclusively. In the case of Apple, it was obvious that Apple intended to force all app purchases through the App Store. If the U.S. Copyright Office had not rejected the argument, Apple would have a monopoly on apps for their machines and could raise prices or block purchases without competitive consequences.

The same applies to OEMs across the technology spectrum. Under the guise of protecting possible violations of IP, OEMs have seen that customers can be bullied into treating machine code as a licensed product but without any of the clarity usually provided in a license. Users are poorly served by this approach, as a real license agreement would state the specific purpose of the product, terms and conditions of use, transfer, and support.

Not only are most licenses for machine code vague, if they exist at all, but most of them appear as check-box licenses that must be agreed to before installation. These "in-the-box" license agreements are provided postsale and are not negotiable. The buyer either clicks "I Accept," or does not complete the installation. The problem for consumers is the lack of opportunity to review and understand the consequences of the agreement far enough in advance of purchase to consider competitive options. Many of these machine code agreements include terms that restrict the buyer from activities, such as how the machine can be used, that the buyer might decide makes their potential purchase less attractive. It is not that these limitations are illegal, but that the purchaser does not have a reasonable opportunity to understand the terms in advance of the purchase.

Business procurement managers should not wait for the OEM to provide disclosure of the terms and conditions for machine code. Each buyer should be asking "What are the terms and conditions governing the use of machine code for this machine" in the same manner that the questions are asked "What is the power utilization of this model" and "What is the warranty period."

Most fixes to machine code are made when the product is new in the marketplace and field use reveals unexpected problems. Over a few months most easily identifiable problems are fixed and the machine code is essentially frozen. Since there are few parts that remain in constant production for more than 2 years, the next iteration of the product will undergo the

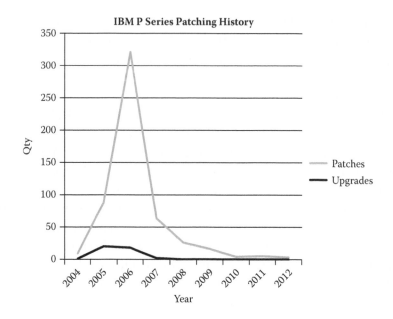

FIGURE 8.2
IBM P series patches and fixes.

same shakeout in the field. Some fixes are applied to an entire production series of products, while others may be specific to particular combinations of products.

Figure 8.2 tracks the history of patches and fixes available on the IBM Fix Central Web site for the P series models beginning in 2004. The top line counts the number of patches and fixes issued for the model and the lower line counts the "Upgrades" that were specifically included in machine code patches at the same time. All upgrades were additions to machine code needed to support new hardware features that became available after the original code was written. The patches were all identified as fixes to known problems, some of which had only specific configuration applicability.

In addition to confirming that machine code upgrades are self-serving changes that allow users to buy more hardware, the chart also proves that the vast majority of patching is done in the first 3 years of model availability. Users considering buying "support" of machine code would do well to scrutinize the patch history of any older product and evaluate if patching is at all active or meaningful.

Manufacturers know that they need to provide access to all patches and fixes so that their equipment problems (defect support) can be resolved as part of both warranty and postwarranty support. The resulting library of machine code options is maintained by the OEM so that technicians and end users can reload code that has been corrupted or lost in the course of time.

Many, if not all, manufacturers currently post their machine code library on a Web site for access by technicians or owners. Often such access is controlled with passwords or validation of support contract status. These types of "security" features serve less to protect the webserver from hacking as it does to support marketing of maintenance agreements. If the OEM is successful in removing access to machine code from owners, then the owner can only retrieve machine code through the OEM, almost always at extra cost.

Such is the power of fear of lost access to machine code that manufacturers have been able to deny access to these bits of code to all but those with active hardware maintenance service contracts or, worse, unrelated software maintenance service contracts. The intention is to compel users to enter into high-margin support contracts exclusively with the manufacturer under the guise that only the technicians hired by the manufacturer are qualified to restore the code already provided as part of the original hardware purchase. This serves to bolster profits to the manufacturer but simultaneously destroys hardware asset value.

Tangible Asset Limbo

The contractual ability to freely transfer hardware inclusive of machine code is the basis of treating technology products as tangible assets instead of licenses. Ownership implies the ability to sell, borrow against, or transfer assets without restriction. Manufacturers that deny owners the free transfer of machine code are in the position to prevent a sale or transfer, thus putting the equipment owner in the position of a licensor or renter, but without the advantages of equipment rental.

Rental of equipment is far different from ownership. If the manufacturer is renting the equipment to the user, the manufacturer remains responsible for insurance, taxes, maintenance, and all financial risk. Renters have none of these responsibilities and do not care how valuable, or worthless, the equipment is upon return. It follows that manufacturers that dictate how, or if, the transfer of machine code is made are interfering with the rights of equipment owners to deal with their equipment as tangible assets.

Providing a generic license for machine code does not resolve the issue unless the license states that the machine code is fully transferrable with the hardware, even if the next user must register his use in order to access the Web site for defect support. Buyers that press for details regarding the types of code covered by the description are likely to be met with blank stares. It does not appear that OEMs are willing to be specific because they would then have to defend the pricing of the license for each and every item. Many of these routines would be extremely hard to defend as something that should be relicensed at a cost that was delivered at no cost with the original purchase. OEM fear of disclosure is the wedge that should be used by buyers to negotiate more sensible terms.

Users should expect OEMs to refuse to provide breakdowns. Just as buyers seeking breakdowns of hardware and software can logically require breakdowns for accounting purposes, buyers can command clarity for the details of the embedded license for the same reason. How would a license patch management system be used without specifics? How would a user be able to explain to the chief financial officer (CFO) that prices for support on an old model machine just jumped significantly?

FUNCTIONS OF MACHINE CODE

See Figure 8.3.

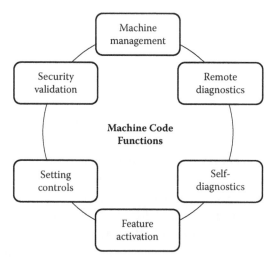

FIGURE 8.3
Functions of machine code.

Machine Management and Access Control

Feature Activation

There are manufacturers that build complex products fully populated with all features, which can be activated at a later date by a software key upon payment by the end user. The routines that control the chargeable feature activation are often lumped together in the definition of "machine code," but the purpose of this code is specific. The makers of these products do not want to permit unauthorized use of these routines for obvious reasons; they intend to be paid to activate chargeable features according to their contract terms. None argue that this code should be under the control of the OEM.

Service Access and Settings Passwords

Buyers of digital electronics are largely unaware that the manufacturer may consider the control of settings within their machine to be owned by the manufacturer and not the buyer. This has been a particular problem with settings that interact with the manufacturers' service teams, such as the call home remote diagnostics options included in many machines.

The call home function is not the marvelous convenience feature that it is sold to be. The remote and hidden trouble call rides outside the owner's repair and service management software. It is impossible for owners of these types of devices to capture management reporting on the failure rate of the equipment, and are thus unable to assure themselves their purchases were the highly reliable devices they were sold.

The other advantage to the OEM is the ability to disconnect the owner from Service Level Agreement (SLA) performance. If the machine (almost always a highly redundant machine) reports a problem behind the scenes, the OEM totally dictates how the parts are provided and the technician scheduled, which could be days, and not hours, as stated in the SLA.

Other setting controls that are common in telephony and networking equipment, are often buried in application software and accessed via maintenance passwords. It is the choice of the manufacturer to supply hardware setting functions within licensed software, and not the choice of the equipment owner. If the licensed application cannot be separated from hardware functions, then there should not even be a purchase option since the hardware is effectively licensed.

Worse, it has also been the case that faced with competition for a service contract, manufacturers have issued patches that remove any opportunity for the equipment owner to access the machine for settings or service. This is the equivalent of an auto dealer sending the auto owner a notice of "Recall" and then failing to disclose that the whole purpose of the recall is to lock the hood and remove the keys.

Maintenance Access Control

When data centers first began, there were many people working inside the data center moving tapes, pulling reports off printers, and so forth. It was some years before most data centers became physically secure and inaccessible except by authorized personnel. In that period, it was obvious that the sensitive parts of machines should be physically locked to prevent accidental or unauthorized access. Whatever physical keys were needed to open access panels were kept handy for the maintenance team, but the keys were the property of the owner.

Fast-forward to today where data centers are often "lights out" and minimally staffed, but the tradition of locks persists. Physical locks have been replaced with access codes and passwords, now claimed as intellectual property, because they are programmable and not physical. This is entirely to the advantage of the manufacturer since the equipment owner is now led to believe that the logical keys are not theirs but rather the property of the manufacturer.

There has been litigation on this subject unfortunately settled out of court. StorageTek (STK), prior to its acquisition by Sun, sued a third-party maintenance provider over "unauthorized" use of the service passwords locking the robotic tape library. The not very secret terms of the settlement are that STK decided to permit this particular maintenance provider unique access to the service passwords, plus coughed up an unspecified sum. STK obviously did not want this conceptual loss to become precedent law.

If one owns a piece of equipment, one certainly should also have all the keys (physical or logical) to access the equipment with or without the permission of the seller. Owners of digital equipment should expect full control of the "keys" just as the buyer of a car takes the keys or the buyer of a house has keys.

Feature Validation

Many vendors use machine code to validate usage of externally licensed products. This is extremely important to the providers of content (media, games, etc.) as antipiracy features, but complicates or prevents repair by the end user. The problem lies with the universality of the block. Legitimate users with no criminal intent are prevented from accessing the machine for repair purposes or unlocking the product for use of products not provided by the original OEM. In many cases, the validation step will lock up the machine ("brick") if unauthorized usage is detected.

Contracts for products that include such features should be negotiated to assure the owner of reasonable rights to use the equipment. An interesting example of the depth of this problem is revealed in a Library of Congress discussion regarding the use of Sony PlayStation 3 (PS3), which was known to be a powerful computing platform, as well as a gaming console.[1] Academic researchers hit upon PS3 as a well-priced and powerful alternative to conventional computers and modified them for use with the Linux operating system. Sony eventually removed the ability to unlock the machine to better protect its game sales and in so doing inadvertently prevented new buyers from converting their machines to other uses. Sony defended its position by arguing that researchers have plenty of other platforms that they could purchase for computing.

Other uses of feature validation within machine code include execution of security and tampering of controls, which are desirable in several industries. For example, the payment card industry (credit cards, debit cards, gift cards, etc.) has developed a set of optional security standards for retailers to follow. Devices marketed as "PCI Compliant" adhere to these recommendations, which often include self-tests for potential tampering. The device is designed to *brick* itself if tampering is detected. This is a useful concept for preventing theft of credit card machines to access the stored records. However, restoration of the bricked device is controlled by software unlocking code, which the OEM may refuse to provide to the equipment owner, despite the owner being at risk for credit card losses and not the OEM. Large users have utilized their buying clout to ensure their right to unlock their own purchases.

Whatever privacy or security fears exist, including very legitimate fears, have to be aligned with the rights of buyers to control their purchases. The current law does not take into consideration the rights of the buyer, so buyers are on their own when it comes to negotiating their rights as they see them. It is currently up to buyers to insist that the piracy fears of the OEM are not their issue to resolve and contract for appropriate access rights to all elements of the hardware purchase.

Diagnostic Routines and Error Reporting

Manually locating errors in electronic circuits is painstaking work. Diagnostic equipment such as oscilloscopes, signal generators, and voltage meters were all essential tools for the customer engineer (CE) to use to identify problems with early versions of electronics. Most IT products today are designed to avoid the need for separate diagnostic equipment and highly trained CEs in the field by being shipped with onboard (embedded) self-diagnostic software.

Many other industries, such as automobiles and aircraft engines, have designed their products to use a wiring harness to connect external diagnostics equipment (themselves computers) in order to capture and process all the signals and data captured during operations. The selection of a diagnostic delivery system is irrelevant, so long as the diagnostic function (including output) is available and defect free.

The output of diagnostic programming is a slightly separate problem for repair in that error reporting and the codes needed to interpret the problems are needed for repair. If the error reporting is not self-evident, such as a dishwasher display that states "check the input value for clogging," then the owner must have access to the meaning of the error codes or be unable to pinpoint the problem. In many industries, such as aircraft electronics, manufacturers deliberately withhold access to error codes to prevent competition for repair.

Any defects to the diagnostic function that are repaired using microcode or firmware patches are just as essential as defect support for other functions. Updates making such corrections (as distinct from upgrades) are necessary to assure all owners that their equipment is running correctly. This includes all levels of diagnostic support, including the most sophisticated. Manufacturers have skirted mandates to provide diagnostics for some products, as are required in the European Union, and instead have provided a simplified version that is available to end users and independent support technicians. The result is that one set of diagnostics is vastly more powerful than the other, leading the OEM to retain a commanding advantage for providing support.

Diagnostics are essential to the equipment repair process both to diagnose the initial problem and then following repair to confirm functionality.

The successful execution of diagnostics is also the standard by which OEMs confirm installation is complete before turning equipment over to the end user and triggering billing. Many OEMs will take used parts under their service contract provided they run diagnostics.

As with other types of machine code, the legal status of diagnostic software is not clear and must be negotiated. Users that have unfettered and clear rights to diagnostic routines (and any associated patches and fixes) are better positioned to retain used equipment value, allow for independent support, and keep their equipment installed and running indefinitely.

Remote Diagnostics

Many machines are shipped with the capability to "phone home" and report problems. OEMs in the enterprise space have commonly used restrictions on remote diagnostic routines to control break–fix maintenance contracts by limiting access to the telephone number directing the call and treating the entire process as proprietary. The OEM does have proprietary rights to their back office service interface, but the OEM is not the only party capable of integrating a machine generated call into a trouble reporting system.

Furthermore, parts replacement, particularly in the era of hot-swappable disk drives, is a physical process that can be performed easily, provided that the location of the failed part is known. End users would find that their equipment would have far greater value and last longer if they demanded control of the call-home telephone number.

EMC is a well-understood example. The EMC machines are designed to run internal diagnostics and report on machine condition, particularly part failure, remotely. With extensive redundancy, the end user does not usually know that a failure has occurred, although EMC is dispatching parts and technicians in the background. EMC does not permit (as of this writing) the call-home system to report failures to the end user nor any other party on behalf of the end user.

End users should also note that remote communication services work both ways. The same network connectivity that allows a machine to dial out also allows an OEM (or other) to dial in to check the machine status. It has been the case that some OEMs have used this dial-in feature to disable functions remotely. End users that have purchased their equipment may want to consider disabling vendor access for the purposes of snooping on the basis of general security policy.

Configuration Access Control

Very commonly, OEMs will provide passwords to engage in configuration or setting "maintenance" functions. This is not the same as hardware break–fix although some OEMs aggregate both services under the heading of "maintenance."

Owners with equipment with configurable settings need access to these maintenance functions for basic utility and long-term value. An example is a telephone switch where maintenance does not mean hardware break–fix but means control of active phone lines and user authentication. Whereas predigital switches may have needed a skilled technician to move lines within a plug board, digital switching has no such linkage. Password-protected access to a control system is a reasonable precaution, but the passwords and control access belongs with the owner and not the OEM under a break–fix agreement. Otherwise the OEM is entirely controlling the use of the device beyond the passage of the title. This would be the same as buying a car, then needing to hire the dealer for eternity to fold the seats up and down in the back to store cargo or use a baby seat.

Feature Activation

Some machines are shipped with features that can be activated on demand. Classic examples are activation of additional onboard processors as computing requirements for power grows. This is more common with products that have per CPU licenses for the systems software and applications software as end users strongly prefer to keep license costs to a minimum wherever possible.

Activation of onboard features is done with machine code routines specific to the purpose. The process of hardware break–fix has no relationship to this code. OEMs may add restrictions on access to feature activations to their machine code agreements, but the impact on the end user (other than those seeking to cheat) is zero.

REPAIR ISSUES

Each manufacturer treats the obligation to provide error correction to machine code differently, but the following are typically standard:

- Safe to operate (no electric shocks or fire risk)
- No spectrum conflicts
- Functions as specified
- Secure (blurry definition)

Corrections to machine code errors can only be provided by the hardware OEM, not because independent programmers could not create patches, but because only the OEM has the design details necessary to identify and make corrections. As such, users are completely dependent on the OEM for such support. Unless an OEM was to offer their machine code as "open source," there is no potential for independent support of machine code whatsoever. Even if open source, few commercial users would want to allow modifications to this most essential code. Individuals are a different story. Many gaming hobbyists tweak machine code to "overclock" their computers; auto enthusiasts "tweak" their engine control modules to change engine timing.

Defect support for machine code is a critical element of hardware break–fix. The end user is often unaware of how policies regarding access to machine code predetermine their choices of break–fix provider, destroy or support equipment value in the secondary market, and govern if they can keep their equipment in use beyond the OEM "warranty" period.

How machine code is treated in international copyright law will determine if products including machine code can be totally owned and resold, or if the presence of machine code on any device allows the manufacturer to require a license agreement for machine code. This is an international issue because trade treaties require participating countries to conform their laws for consistency. Whatever copyright changes are made in the European Union or the United States will eventually become part of such treaties.

Rather than start with a discussion of how machine code should not be covered by copyright law (or at least not current copyright restrictions), this section begins with the arguments being put forward in support of copyright protections for machine code (also known as embedded code).

1. Machine code is programming.
2. Authors of machine code are entitled to protection from unauthorized distribution.
3. Support of machine code is an ongoing effort and benefits the owner.
4. Programming is the heart of innovation.

Machine code is undoubtedly programming, but so is all the programming that is delivered as part of any semiconductor chip. There is no functional distinction for the user. Code that controls logic on the chip can also be written to be accessed dynamically as a stored program. Manufacturers shift back and forth frequently between machine code that is fixed into a chip, corrected (updated) in the field with downloaded patches and fixes, and back into a completely physical form into a newer version of the chip.

It is also the case that the hardware design team writes the machine code and supports it. If there is a flaw in the design of the onboard programming, the hardware engineers diagnose the problem and issues patches and fixes if the problem can be fixed. If the product is impossible to patch even temporarily, the OEM may be faced with providing a physical replacement version of the part or product. Change levels at the machine code level are commonly treated as revision levels or engineering change levels. It has been common for OEMs to make all such fixes available to all buyers regardless of warranty status in order to make sure their reputations are not diminished. OEMs do not want to be faced with customer backlash over being denied access to patches and fixes correcting serious design flaws.

The vast majority (certainly by volume) of digitally driven products are manufactured without any update capability for machine code. The chips that are used in microwave ovens, programmable coffee pots, and televisions cannot be patched externally (unless they are networked). These chips all include machine code, but the code is so static that millions of users have operated the machine without any concern about bugs in the machine code. Nor do users have any restrictions on resale at garage sales (or auction sites) based on copyright claims to machine code. The absence of an update capability is not protection for buyers of any digital electronic products as there are no laws that protect owners from being treated as users of intellectual property if the OEM decides to make such a claim.

Copying of machine code is not a likely profitable activity for pirates. The only buyers for a pirated copy of a machine code routine would already have a copy installed on the part for which the code was written. If aware, owners of equipment would not repurchase that which they already have. Nor does machine code transfer readily to other machines as it is hardware specific and is often configuration specific. Although the OEM may reuse bits of machine code from model to model, the end user has no reason to pirate versions of machine code for machines they do not own.

The potential of a black market in machine code seems unlikely given the naturally limited market for code that is already provided with the machine. However, if an OEM were to block access to machine code to existing machine owners, there is a potential for users to seek replacement copies elsewhere. The vast majority of commercial IT managers and executives would never introduce pirated copies of system-level code into their environment. The career risk of a code failure taking down a system would deter all but the most idiotic.

Another logical nail in the coffin of worries about proliferation of machine code is the problem of the first sale (not the legal term but literally the initial sale). The vast majority of technology products are sold with no reference at all to the existence of machine code. Even when manufacturers change policy and add a requirement for a license for machine code, the manufacturer theoretically cannot go back in time and create restrictions that did not exist at the time of the original sale.

Unfortunately, not all users or counsel picked up on what may have appeared as "boilerplate" terms granting the OEM the right to change terms unilaterally in the future. It is clearly in the best interest of contract managers to review all old agreements and make sure they did not accidentally make such agreements. If part of master purchase agreements, new agreements can be required as part of new business, and any of these claw-back provisions should be removed.

The thorniest of issues involving machine code is not the initial provision of machine code but corrections to machine code errors. As discussed in the section on defect support, the OEM does not repair code that is working properly, so the updates provided to machine code must be regarded as important to the operation of the machine. The manner in which equipment owners are allowed to access these code updates determines if the owner controls the machine code as part of the equipment purchase or if the OEM controls the use of the machine beyond the initial sale.

There is no doubt that updates to repair flaws in machine code are a necessary service to buyers. Buyers undoubtedly want their purchased equipment to operate as specified. The question is if machine code corrections (updates) are distributed freely to owners as part of the vendor's view of its responsibility for defect support or if the vendor prefers to view corrections as chargeable. Until a determination is made, if it is ever made, in law, end users need to negotiate for how machine code updates are provided or face almost certain losses in equipment value.

The final argument favoring treatment of machine code as licensed and chargeable software is the innovation argument. There is tremendous innovation provided through software, but the flaw in the argument is that machine code is rarely (if ever) the source of the innovation. Machine code is a very limited set of instructions intended for the software developer to use to innovate. It is more like a programming language than a program. Instructions are added to the machine instructions for models of machines to access and use new features—and hardware features at that.

Just as not all Java APIs are copyrightable,[2] not all machine code is either. Machine code is not creative in its own right. An elegant and tightly coded routine saves processor cycles for executing the higher-level instruction, but does not provide a new way to image the universe. Contracts that treat machine code as part of the hardware do not change the incentive for innovation on the part of the manufacturer. OEMs are still improving their products, adding new features, and making more powerful machines to drive new sales. Machine code is written to allow new features and products to operate, not to be the innovation.

CONSUMER ISSUES

Consumers are currently prevented, or thwarted, from self-repair on a growing variety of products based on the presence of machine code/embedded code on digital parts. In the case of consumer items, machine code is not only used for hardware patches and fixes, but commonly to lock the unit from tampering and to validate the presence of authorized parts and software as part of the operation of the unit. For example, gaming consoles validate the legitimacy of the game license through machine code. Some products are set to lock everything (brick) when nonauthorized products are detected. This use of machine code is for purposes of antipiracy. (See Figure 8.4.)

At the physical level, there is no technical distinction between consumer products and business class products. These are marketplace differentiators. Unfortunately, the U.S. Copyright Code is set up to deal separately with different "classes" of equipment as though they are completely different products, which they are not. The same technology in a cell phone is also in a server and is also in a combine harvester.

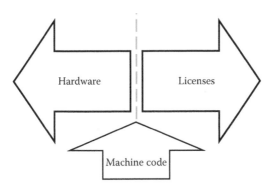

FIGURE 8.4
Machine code divides hardware from licenses.

The following applies if the OEM treats machine code as intellectual property:

- Break–fix—Machine code patches are the primary delivery vehicle for error correction in hardware. OEMs may use their unique ability to deliver patches to pressure end users to purchase all postwarranty repair services in order to have access to updated machine code. Machine code is commonly used in consumer products to lock products.
- Secondary market—If machine code and patches are not transferrable to a secondary owner, then the equipment is effectively worthless. There may be some part value, but the functional machine will not be attractive to a used buyer since it cannot be operated. Consider the auto analogy that a vehicle cannot be sold as a used vehicle because the keys to the ignition are not transferrable.
- Sweating assets—From the preceding sections, it is clear that OEMs have a strong financial incentive to imply that machine code is IP because it creates the situation where they can monopolize the lucrative market for break–fix service and at the same time compel users to purchase new models as it suits the revenue goals of the OEM.

If the OEM treats machine code as hardware, then the following policies apply:

- Break–fix—Owners can easily download whatever patches and fixes apply to their specific purchase for purposes of repair. Owners can hire technicians of their choice or perform their own repair without restriction.

- Secondary market—Buyers of used machines can do the same, facilitating the reuse of equipment and the preservation of investor value.
- Sweating assets—Owners of technology products can continue to use older models until it suits them.

―――――――

COPYRIGHT ISSUES: DIGITAL MILLENNIUM COPYRIGHT ACT AND WORLD INTELLECTUAL PROPERTY ORGANIZATION

In the late 1990s, Congress did a major rewrite of the copyright code not only because the prior code did not deal well with the growing electronics industry but also because reform was necessary to bring the U.S. government into compliance with international trade agreements, notably the Berne Convention and the World Intellectual Property Organization (WIPO). The provisions of the copyright code that were most lacking in international terms had to do with anti-circumvention, which was included in the WIPO Treaty of 1996 and adopted in the United States through the Digital Millennium Copyright Act of 1998 (DMCA).[3]

The section of the DMCA that adopted WIPO terms regarding anti-circumvention is Section 1201. The language is clearly directed at the potential for users to purchase products (both hardware and software) that would facilitate copying of movies sold using VCR technology. The unfortunate consequence of the language is that it was so broadly drawn that the same language has been used to litigate against all vendors of technology tools including those needed for basic repair. This clause is known as the *anti-circumvention* clause of the DMCA.

(2) No person shall manufacture, import, offer to the public, provide, or otherwise traffic in any technology, product, service, device, component, or part thereof, that—(A) is primarily designed or produced for the purpose of circumventing a technological measure that effectively controls access to a work protected under this title; (B) has only limited commercially significant purpose or use other than to circumvent a technological measure that effectively controls access to a work protected under this title; or (C) is marketed by that person or another acting in concert with that person with that person's knowledge for use in circumventing a technological measure that effectively controls access to a work protected under this title.

Under Section 1201, digital electronics manufacturers have effectively prevented independent development of alternative forms of diagnostic software, diagnostic hardware, and any tools useful to access software locks of any kind, including for equipment deemed obsolete by the vendor. Most discussion about "copyright reform" in the news centers around this section, because of the consequences of how the section has been interpreted to allow manufacturers to establish unanticipated monopolies for support, far outside the original intention. It is clear that the intended protection of creators of media content from having their works proliferated has been taken to legitimize the total blockage of access to machine for any purpose.

The DMCA has another section that relates directly to the repair of technology products but is barely applicable despite its origins as the Computer Maintenance Competition Assurance Act.[4] Now known as Section 117, the section has a paragraph quoted below in specifically granting the end user the right to backup and restore copies of licensed software in the event of a repair. At the time this section was negotiated, framers were aware that this section was not entirely a right to repair and that it might need to be amended to better clarify this point.[5]

(c) Machine Maintenance or Repair—Notwithstanding the provisions of section 106, it is not an infringement for the owner or lessee of a machine to make or authorize the making of a copy of a computer program if such copy is made solely by virtue of the activation of a machine that lawfully contains an authorized copy of the computer program, for purposes only of maintenance or repair of that machine, if—

(1) such new copy is used in no other manner and is destroyed immediately after the maintenance or repair is completed; and

(2) with respect to any computer program or part thereof that is not necessary for that machine to be activated, such program or part thereof is not accessed or used other than to make such new copy by virtue of the activation of the machine.

The status of machine code/embedded software in copyright law is one of the major conceptual legal questions facing DMCA reformers. Part of the problem is that machine code is not mentioned anywhere in the U.S. Copyright Code, specifically the DMCA. This is almost certainly because the routines we have come to define as machine code were being treated entirely as hardware at the time the DMCA was being thrashed about in

Congress. This theory is further supported by special (*sui generis*) protections created at the time for the manufacturer of semiconductor circuits to grant the masks (essentially a form of printing) copyright protection separate from patent protection.

In the absence of specific mention of machine code, everyone is left to wonder if copyright law or patent law applies. How machine code is to be treated is a core issue for Congress to decide, if and when it takes up revisions to the DMCA. In the meantime, users and vendors will argue and negotiate directly. In the absence of settled law, or even a body of precedents, users have tremendous flexibility in how they decide to deal with such issues.

Copyrightable Code

One of the reasons that litigation is often settled out of court (see the next section) is that not all vendor claims of copyright infringement are accepted by judges as appropriate. Copyrights are filed by vendors for just about everything, but that does not mean the material would be judged "copyrightable" in a court challenge. There are standards of originality and specificity that have led to rejection in the courts. Often, when pressed, vendors will withdraw weak claims rather than attempt to justify them in front of a jury.

For purposes of machine code, the first question is one of originality. If a series of instructions is simply the most obvious way to get data from point A to point B, then the code is not likely to be copyrightable. This is the same reason that a telephone directory is not copyrightable: a listing of names alphabetically is not original. Many user manuals might not be sufficiently original to provide copyright protection for copying. There have been cases where judges ruled that password protection routines are not sufficiently novel to be granted copyright protection.[6]

The second requirement is one of *specificity*. In *Oracle v. Google*, the judge ruled that the application programming interface (API) as a concept was not specific enough to warrant copyright protection. The ruling stated that only an individual API may be copyrighted. Despite the enormity of this ruling, there are not yet enough court rulings to treat these issues as *settled*, so users should not expect corporate counsel to find lots of precedents to follow.

LITIGATION AND LEGISLATION

Copyrights are different than patents in that they protect the rights of authors (most specifically creative works such as paintings, music, and literature) from being proliferated without the permission of the creator. Patents protect the manufacturer from having their innovations manufactured and sold without their permission. The capital costs to manufacture innovations in electronics have proven to be an effective protection for patents, whereas the Internet has made the proliferation of copyrighted materials simple enough for toddlers to violate.

As a result, most litigation related to software and content is centered on copyright violations. Even litigation over hardware-centric routines, such as APIs, has been pursued through copyright law, as we saw with *Oracle v. Google*.

End-User Litigation

It is a basic sales principle that vendors seeking to write new business in accounts do not "bite off the hand that feeds them." This is a point of weakness in vendor negotiation that is poorly understood. Most users, and most general counsel for users, try to avoid having their company involved in any hint of dispute. This is a public relations issue and not a technical one. Users can file complaints of their own in a variety of settings, including the most powerful—the court of public opinion. The challenge for users is to be willing to take the backlash from the vendor.

The most significant threat a vendor makes to end users in order to forestall being publically embarrassed is the threat of loss of "support." Support means many things, not all of which are equally threatening. Only users, with corporate counsel, can decide if the threats are realistic or would evaporate if exposed.[7] Vendors are not always expert poker players and can be made to back down if the user is inclined to put up a fight.

Competitive Attacks

Vendors are more likely to focus litigation against competitive suppliers and service organizations as better targets. Well-heeled manufacturers with in-house litigators can engage in almost unlimited legal assaults on competitors even when the assaults are later dropped. The mere action of filing a lawsuit is quickly spun to end users as a warning that the competitor has done something awful. The competitor must fight back but keep business going at the same time. Fighting for new business, fighting a legal battle, and running a business at the same time is too much for all but the largest companies to undertake. As a result, most cases are settled out of court.

Out-of-court settlements leave little precedent law behind for users or competitors to turn to for guidance. This leaves users again in the position of needing to negotiate even the smallest of potential issues before signing the purchase order.

Legislative Efforts

Most of the legislative advocacy work being done on intellectual property issues is happening at the behest of consumers and may result in DMCA reform at a later date. The aftermarket auto industry has successfully passed legislation in Massachusetts giving consumers in that state specific rights to repair that are fundamentally similar to digital repair, but the statute was specific to automobiles. There are at least two consortiums pushing more generalized rights to repair: (1) a federally focused group under the mantle of the Owners Rights Initiative and (2) a state-focused organization pushing digital right to repair at the state level.[8] There is also legislation pending in the House known as the Promoting Automotive Repair, Trade, and Sales (PARTS) Act that is focused primarily on automotive parts but is getting close on more general digital issues.

The most likely first revision to the DMCA will be cell phone unlocking due to a vast grassroots rebellion that emerged as soon as unlocking was made illegal again by the Library of Congress in 2013. Callers overwhelmed the White House, which suddenly found interest in unlocking. Several bills were drafted within weeks and were in front of the entire Congress is less than 6 months. Cell phone companies have been fighting unlocking because it would allow consumers far easier access to price-shopping wireless service than with a locked phone.

Objections to unlocking machine code in general are rooted in the fear that any device, once unlocked, is wide open for the user (consumer) to defeat encryption measures that control forms of copyright infringement that the OEM does not want to allow. Most of these objections are coming from the content and media industry, where widespread piracy of movies and music are major legitimate issues. The question for legislators when they go to update or revise the DMCA is how to balance the needs of the consumer to control their purchases, and the needs of content and media vendors to make it more difficult to be pirated. The current situation is clearly out of balance, since consumers are unable in many cases to make reasonable repairs to their purchases—a situation about which the content vendors are aware and not conceptually against.[9]

As with help from court rulings, users should not plan on any legislative assistance in protecting their rights in the near term. As with all legislation, the devil is in the details. Moving bills from concept to passage is an exercise in making sausage; the results are not always pretty and may never resemble the original intent. Bills that address issues within a particular state may prompt buyers of equipment to prefer doing business in that state but will not address either copyright or patent law issues directly because those are specifically federally controlled.

It is currently the case that OEMs are making very generalized claims of copyright related to machine code, which may be very weak given the above. Not surprisingly, the vast majority of buyers are not looking to fight vendors over nuances of copyright law and will simply make their purchases accepting OEM claims that machine code is subject to copyright law. This is lazy.

Savvy buyers can change the paradigm of negotiations for access to machine code by requesting clarification regarding the specific copyrights for all routines they are being asked to treat as copyrighted. It is very likely that many copyrights have been filed, but the specifics may reveal that the routines are of limited value. For example, if a copyright were filed for machine code that added support for a new product model, a buyer of a different model might not agree that the OEM should charge for access to updates providing such self-serving code. It is only by having details that any negotiation can be fully complete. The worst case for buyers is that the code purchase is supported by detail. The best case is that the proposed charges are waived.

SUMMARY

Buyers who focus on how machine code (and synonyms) is treated in their agreements will reap rewards in lower total costs of ownership as they can sweat their assets far beyond the intentions of the OEM. The legal environment is not settled, giving current buyers far more flexibility on how to proceed in negotiations than is presently undertaken.

NOTES

1. See testimony of May 17, 2012, at UCLA regarding Section 3, beginning on p. 8, http://www.copyright.gov/1201/2012/hearings/transcripts/hearing-05-17-2012.pdf.
2. Oracle sued Google for $6 billion in patent and copyright violations over use of Java APIs. The court ruled in favor of Google, handing Oracle a massive defeat. For details, see "Oracle v. Google," *Wikipedia*, http://en.wikipedia.org/wiki/Oracle_v._Google.
3. See "WIPO Copyright Treaty," *Wikipedia*, http://en.wikipedia.org/wiki/WIPO_Copyright_Treaty.
4. See history of the DMCA and various sections at "Digital Millennium Copyright ACT," *Wikipedia*, http://en.wikipedia.org/wiki/Digital_Millennium_Copyright_Act#Title_III:_Computer_Maintenance_Competition_Assurance_Act.
5. The entire text of the section can be found at "17 USC § 117—Limitations on Exclusive Rights: Computer Programs," Legal Information Institute, http://www.law.cornell.edu/uscode/text/17/117.
6. Judge Irenas upheld previous rulings against AVAYA in the case of *AVAYA v. Continuant* as part of a series of pretrial requests for summary judgment. For the entire ruling, see "Avaya Inc. v. Telecom Labs, Inc. et al.—Document 482," Justia U.S. Law, http://law.justia.com/cases/federal/district-courts/new-jersey/njdce/1:2006cv02490/273236/482.
7. Having been through several instances where a vendor "walked off" the account, I can attest that losing the presence of the sales rep is not a bad thing. The equipment kept running, the CE kept working, and the rep eventually returned humbled and cooperative. Users that take threats of being abandoned seriously are losing a powerful weapon: ignoring the vendor.
8. Digital Right to Repair Coalition at: www.digitalrighttorepair.org.
9. Extensive discussion between different sides of the issue is available on the copyright.gov/1201 section of the Web site. The question of repair arose and was elaborated in following the meetings. Letters provided by both Andrew "Bunnie" Huang, PhD, (http://www.copyright.gov/1201/2012/responses/andrew_huang_response_letter_regarding_exemption_3.pdf) and Christian Genetski of the Entertainment Software Association (ESA) (http://www.copyright.gov/1201/2012/responses/esa_response_letter_regarding_exemption_3.pdf) confirms that the ESA understands that hacking a console for hardware interoperability (repair) is not the same as for software (content); see p. 3. No solution was proposed and the question of treatment of jailbreaking for purposes of repair was not taken up in other hearings.

9

Service Parts

INTRODUCTION

This chapter discusses the various sources of service parts available in the marketplace along with a discussion of the types of parts, the different business interests of parts suppliers, their specialties, and limitations. Also included is a discussion of counterfeit and gray market parts, their sources, and risks of use.

PROPRIETARY PARTS

A wide variety of equipment is made for a single vendor or type of function. The designs of these parts are proprietary. The vendor typically does not sell the rights to manufacturer parts using the design, so the only source for the part or machine is the vendor. For example, the Sony Betamax was the technically superior video recording media of its day. Sony chose not to keep the design proprietary and did not license it for production by others, thus keeping the price high. Competitive VCR technology was widely licensed, and competition and volume quickly drove the costs down so rapidly that the purchases of VCR devices exploded. Sony won the technology war but lost the consumer base.

From a repair perspective, access to proprietary parts is often limited by the vendor to its authorized repair channels. Early in the product distribution lifecycle, this gives the original equipment manufacturer (OEM) dominant control over the pricing and availability of repair, but as used products become available, scavenged parts end the parts monopoly. Patent laws protect the rights of the vendor to refuse to allow "generic" versions of their proprietary parts, which tend to keep parts in limited supply but does not prevent repair in the same manner as limitations on intellectual property.

COMMODITY PARTS

Most electronic hardware parts are commodities, acquired and assembled on production lines all over the world. Printed circuit boards may be customized as a substrate to fit various ergonomic or shape requirements, but the parts are still commodity items. Excellent examples of the assemblages of commodity parts within brand name devices are easily found on the iFixit Web site, famous for its teardowns.[1]

The 2012 iPad test unit has the following hardware (Figure 9.1):

- A5X SoC (Samsung made, Apple branded)
- 1GHz dual-core ARM Cortex A9 CPU
- 200MHz quad-core Imagination Technologies PowerVR SGX543MP4+ GPU
- 9.7″ LED-backlit retina display (2,048 × 1,536 pixels at 264ppi)
- 3.7V 43.0WHr 11,560mAh Li-ion Polymer Battery (Model: A1389)
- 16GB Hynix H2DRDG8UD1MYR NAND Flash

FIGURE 9.1
Apple iPad 3 parts teardown. (From iFixit. With permission.)

- Unknown Apple chip (343S0561-A1 12058HCA)
- 1GB Samsung DRAM (512MB mobile DRAM K3PE4E400E XGC1 x2)
- Broadcom BCM4330 802.11a/b/g/n MAC/Baseband/Radio with Integrated Bluetooth 4.0+HS & FM Transceiver
- Fairchild FDMC 6683
- Texas Instruments CD3240 driver device (CD3240B0 1CAY7KT)
- Broadcom BCM5973 I/O controller (BCM59731A1 KUFBG HE1202 P11 179034 03 W)
- Broadcom BCM5974 microprocessor (BCM5974 CKFBGH HE1205 P12 184595 N3 W)
- Qualcomm MDM9600—3G and 4G wireless modem
- Avago A7792
- Unknown MT chip (2CDI8 NQ312 BDDT)
- Qualcomm RTR8600 multiband/mode RF transceiver for LTE bands
- Triquint TQM7M5013 quadband linear power amplifier module
- Unknown Apple chip (338S0987 B0RJ1152 SGP)
- Qualcomm PM8028 Power Management IC

Some parts are Apple designs and presumably *proprietary* but are combined with parts built and designed by others. In the case of proprietary parts, Apple is the holder of the patent for the part and is the only source for that part unless they license production to others. The owner usually controls repair of proprietary parts, but swapping broken parts with new ones is not proprietary. The challenge for repair providers is access to a supply of proprietary parts for service along with whatever associated machine code that might need to be restored.

Within the assembled unit, each of the parts manufacturers is responsible for its own defect support, which is in turn distributed by Apple as the assembler. Defect support also includes the associated machine code, which is the responsibility of the parts manufacturer to correct.

FRAND/RAND Parts

Within the world of hardware patents is a category of parts known as FRAND (fair, reasonable, and non-discriminatory) or RAND (reasonable and non-discriminatory) parts. These are parts designed by Standards Setting Organizations (SSOs) to enhance interoperability and consistency of design for the buying public, such as a USB connector. The design of the USB was intended to be used by many vendors. FRAND/RAND products are sometimes licensed, with royalties. Many of the parts we consider *commodity* parts are in fact covered by such royalty and licensing arrangements.[2]

For buyers of technology parts, FRAND and RAND licensing are not a direct consideration but the production of service parts and the limitations made by OEMs with respect to distribution of service parts are linked to these agreements. For example, Apple does not license the production of glass for the iPhone 5 so the only replacement option is completely under its pricing and distribution. Unless Apple agrees to license the production of their patented products and parts, the only resource other than Apple for such parts is to scavenge parts from other units.

Original Parts

The OEM parts counter is not the only source of the same parts used by the OEM, often called "Original" parts. Many products come into the supply chain for assembly, such as hard drives, and are easily located on the open market as the identical part. OEMs occasionally make deeply discounted volume sales agreements with distributors to move inventory, which are then resold as whole machines or stripped for parts. As equipment is replaced, older models are sold in the secondary market and frequently stripped for parts. All such parts are *original* but not necessarily new. Most parts sales, including those from the OEM parts desk, are not separately warrantied beyond 90 days, as most parts are quickly integrated into a base machine, which is covered by a service contract associated with the serial number.

The only objective standard for bringing replacement parts under the machine warranty is that the part (or the machine) properly executes all diagnostic routines. There is no evidence that the OEM can predict the failure rate of any part or device with sufficient accuracy to separate new, used, or "remanufactured" parts. Some OEMs have edicts that require the equipment run for a period of time, such as 30 or 90 days, before taking the item under their maintenance contract. This policy provides nothing of value to the customer and serves only to bully clients into buying parts only from the OEM.

OEMs themselves will buy parts off the used market to support their break–fix contracts. These parts are identical to all other used parts but have been *blessed* by the OEM and occasionally represented as "Factory Remanufactured." In the case of electronics, factory remanufactured is a dubious advantage since wear and tear are not obvious on printed circuit boards (PCBs) (excepting bulging or leaking capacitors), the item's prior history use is unknown, and boards are not repaired without having first failed. Cases may have been cleaned of debris and wiped down to appear new, but the innards are not improved.

Plug Compatible Parts

Many parts are built by suppliers other than the OEM to attach readily to the OEM machine using the OEM standard plug or interface cable. A commonly used type of "Plug Compatible" part is add-on memory, which fits neatly into the socket provided by the OEM but is not made by the OEM. It is logical that the warranty coverage for plug compatible products must be made by the compatible vendor, yet some OEMs attempt to prevent the use of any plug compatible parts by voiding all warranty if such parts are used. In an automobile analogy, a vehicle owner would void the entire vehicle warranty by using a generic air filter instead of the factory brand.

Unless the non-OEM part can be shown to have damaged the OEM equipment, owners should be free to add any equipment of their choice for any purpose as a basic right of ownership. This does not mean that the OEM warranty will cover these items. Buyers of third-party memory, batteries, disk drives, or power supplies, as examples, should be able to arrange for support for such parts separately without the approval or interference of the OEM.

Remanufactured Parts

There are some parts labeled "Remanufactured" by the OEM. This does not mean the product has been sent back to the factory for revision or adjustment. It is most often the case that a part that has been returned as a failed part is repaired and thoroughly tested and returned to inventory as a spare. Some OEMs call such items "Factory Refurbished" or "Remanufactured" even though the process of repair and testing is identical when performed by technicians anywhere in the world. Electronics are repaired and returned to service, or they are trash. According to specialist companies performing board-level repair for OEMs and others, roughly 95% of electronics can be repaired and returned to service.

Not all machines can be returned to service with a little dusting on the inside and smudge removal on the outside. In the case of machines subject to physical wear and tear, such as high-speed laser printers, the factory refurbished or remanufacturer nomenclature has legitimate meaning.

Rebuilt machines are restored to original specifications which would have worn parts and degraded performance.

Refurbished Machines

The "refurb" process for machines is not the same as remanufacturing. Refurbishing is largely a cosmetic process; sometimes more of an auto body shop than a technology process. If the machine is dented or damaged, the refurbisher can straighten dents and repaint frames. Dust is blown out of cases and keyboards; smudges and other signs of wear cleaned or polished out. Refurbishing is used primarily to improve the appearance of used equipment prior to being delivered to a secondary market buyer.

GRAY MARKET PARTS

The wording of "Gray" market parts was created by OEMs as a deliberately pejorative term intended to hint at some nefarious association with black market parts. We are left to presume that black market parts would be illegal parts, however that might be defined. These same "gray" products could be just as easily labeled blue or chartreuse parts because they are all legitimate OEM parts sold in the same way that overstocked merchandise is traded around the world. One of the largest sources of whole machines sold as gray market is the OEM selling merchandise at different price points either overseas or through discount channels.

OEMs do not like to admit that they are in control of their gray markets. As shown in Figure 9.2, the OEM is in control of most of the potential sources of both gray market (legitimate surplus product) as well as counterfeit sources. The opportunities for economic benefit on a large scale lie in the sourcing and assembly of products, not the postdistribution disassembly of finished products.

Whenever an OEM sells equipment in countries at different prices, it is creating the value differential that encourages dealers to acquire the product overseas and import it. This is not illegal nor does it harm the product. This was recently affirmed by the Supreme Court decision *Kirtsaeng v. John Wiley & Sons, Inc.*, which appeared to be a case about importation of textbooks but was ultimately a case about the purchase and importation of copyrighted material (including equipment) acquired overseas.[3]

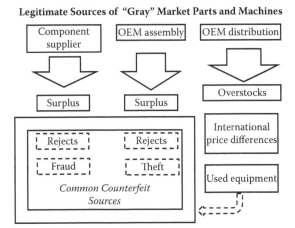

FIGURE 9.2
Sources of gray market and counterfeit parts.

Another major source of gray market equipment comes from OEMs dumping excess inventory through distributors at deeply discounted prices in much the same way that overstock merchandise is traded for all other goods. OEMs that complain they have lost profit due to gray market sales are not due any sympathy. They made the deals at the prices they wanted that resulted in the products being shipped and they booked their revenue accordingly.

Gray market equipment is under attack by an alliance of OEMs named AGMA, standing for the Anti-Gray Market Alliance. The AGMA Web site (http://www.agmaglobal.org/cms/) makes a number of weak assertions about gray market equipment, already discussed, and one very large and untrue allegation: that the gray market is the primary source of counterfeit equipment.

COUNTERFEIT PARTS

There are always frauds and cheats in any industry. Electronic equipment is no exception. There are manufacturers, assemblers, and packagers that insert counterfeit parts without the knowledge of the OEM. There are traders who will knowingly buy counterfeit parts just as some pawn shops trade in stolen property. These are criminal enterprises and cannot be countenanced. If it were true that counterfeit equipment is flowing into the marketplace solely through the gray market channel, none of us would expect to buy legitimate brand merchandise at Costco or Overstock.com.

Counterfeiting electronics is not a problem of completely fake products. Just as a fake Gucci handbag is an actual handbag, counterfeit parts not only look like the OEM part, they also operate. (If the parts did not work the machine would not be sold or would be almost immediately returned for credit.) Since the products must work, they must come from some other vendor than the vendor that was specified. This is most likely done during the assembly process where the greatest opportunity for part substitution occurs with the greatest opportunity for illegal profits.

From a security standpoint, there is the potential of parts being inserted as a type of Trojan horse for spying or mischief. The most common concern is more ordinary: expected performance and durability. Most likely, counterfeit parts are just cheaper and someone in the legitimate supply chain made a tidy personal profit from the switch. It is also known that equipment rejected as substandard during quality checking in production runs may be diverted into sales instead of scrap.

The following from the U.S. Department of Commerce, Office of Technology Evaluation (January 2010)[4] succinctly explains just how difficult a task it is at the point of assembly to discern gray market parts from normal parts:

> Circuit board assemblers, as integrators of electronic components, are an "invisible" part of the supply chain. They are not manufacturers of electronic parts and components, nor are they end users of the final assembled circuit boards. These companies are intermediaries, and as such do not always experience the negative effects that counterfeit parts have on the rest of the supply chain. Circuit board assemblers generally do not focus on testing for counterfeits, often assuming that their suppliers have provided them with the proper parts. When they do conduct testing for counterfeits, it is typically a form of visual inspection on the incoming parts.
>
> While completed circuit boards are tested after assembly, this testing is usually for performance. In fact, most assemblers discovered counterfeits through customer returns. Although assemblers encountered relatively few counterfeits, their low levels of testing, documentation, and auditing may obscure the extent of the problem.

The same report also clarifies that the vast majority of counterfeit items are discovered only after a part has failed and is inspected, leading to the inevitable conclusions that millions (or more) of counterfeit parts are in circulation and in use—undetected. The extent of counterfeiting is actually unknown. Incidences of counterfeit might range from an unauthorized substitution of a lower cost part at the point of assembly to outright fakes of whole products.

Clearly, OEMs could reduce their vulnerability to counterfeiting by more management supervision and testing during assembly, more domestic manufacturing, better validation and packaging at the point of origin, and offering a comparable validation service for end users or dealers concerned about products domestically.

Thus, end users in ordinary information technology (IT) settings need to understand the real risks of counterfeit items in their enterprises. There are five important facts to review as the basis for the analysis.

1. As much as 10% of all electronics are counterfeit.
2. Counterfeit electronics are present throughout the supply chain.
3. Counterfeits are most often detected when they fail.
4. The majority of counterfeit items are of less than $100 unit value.
5. The vast majority of all electronics are manufactured overseas.

For end users, the problems of using electronic counterfeits are similar to those of buying any other: the product is inferior, the product is not repairable, or the product is a Trojan horse for some nefarious purpose. The Trojan horse scenario is a frightening possibility for parts in defense, intelligence, and critical infrastructure, but not a realistic threat for the vast variety of common business computing.

The important issue for ordinary end users is how to assure themselves that the products they purchased are the quality they expected. An element of the quality concern is that the part be eligible for service/repair contracts in the event of a failure. Assessing product legitimacy before accepting inventory is the number one recommendation of leading trade organizations representing all aspects of the electronics industry. This recommendation is shared by interests as varied as those of manufacturers of basic electronic components such as NEMA (National Electrical Manufacturers Association)[5] to secondary market broker/dealers and independent distribution channels represented by ASCDI/NATD (Association of Service and Computer Dealers International and the North American Association of Telecommunications Dealers).[6] All of these organizations and their memberships are committed to avoiding trade in counterfeit products and have taken steps to assure their customers of legitimacy and proper support.

End users should insist that visual inspection and testing take place ahead of installation so that any suspicious parts can be returned immediately to the source for refund and reporting. Most end users rapidly cull their vendor lists of any incidences of counterfeit shipments. As quality vendors largely test and inspect equipment prior to shipment, the incidence of counterfeit substitution is more common with parts that are "Drop Shipped." End users should try to avoid situations where the product itself is not delivered directly from the vendor, as extra time is needed to do proper testing.

Most quality vendors already protect themselves from the potential negative customer relationship backlash by conducting their own preinventory testing upon receipt of wholesale shipments. In turn, counterfeit items are discovered and removed from the supply chain, and the offending wholesale vendor is usually stricken from the wholesale vendor list.

In an ideal world, no counterfeit part would make it past the scrutiny of the first examination after importation of the product into the United States from Asia. However, by the time an assembled machine is packaged and shipped from Asia, the largest opportunity to prevent counterfeits has been lost. The problem for manufacturers is one of scale and cost-effectiveness. Billions of parts are made all over the world and assembled by the millions in Asian factories. Not all factories or original component manufacturers (OCMs) adhere to the same security, scrutiny, or ethical standards. It has been identified by the Federal Bureau of Investigation (FBI) and U.S. Department of Commerce that a hefty chunk, perhaps as much as 50%, of all counterfeit products enter the supply chain through Asian OCMs and are assembled, unwittingly, into OEM products. Once in the legitimate supply chain for the OEM, the product rides into the OEM distribution system with perfect paperwork, packaging, and the blessing of the OEM.

Few whole machines are fully counterfeit for obvious reasons. Anything that did not work at installation would be quickly rejected. For the counterfeiter to make a profit, the *play* for the counterfeit market is to build lower spec parts or scavenge rejected parts, then substitute them into the supply chain. The result is a machine that works long enough to be installed where it may, or may not, ever be discovered.

This speaks directly to the problem of quality. If substandard parts remain fully functional, the impact on end users is zero. (The manufacturer has an internal problem policing its supply chain, but the end user is not harmed.) Trouble occurs when the substandard part fails, at which point a service technician is called upon to pull and replace the failed item and discovers the counterfeit. The assumption is that the fake item had a higher failure rate than the OEM design, which is the real impact of the use of counterfeits for end users.

Repair organizations, regardless of technician, are the eyes and ears of both the OEM and the end user with respect to identifying counterfeit parts and products. OEM part warranties are short, often 1 year or less, so the OEM itself loses control of the failure rate of its assembled products rather quickly. The better the fake, presumably the longer it will last, so the independent service organization plays an important part of the identification and prosecution of counterfeiters.

End users should be cognizant that no supply chain is pure. Reliance upon the OEM distribution channel will result in an excellent chain of paperwork but still has risk. The best protection for end users is to buy from proven vendors with their own test and evaluation facilities. Unless there is a parts emergency, allow 2 to 3 additional days for each item to be properly vetted before delivery. Avoid the temptation to purchase goods at prices that are *too good to be true.*

End users should also be aware that the OEM has a vested interest in painting nonauthorized providers in a negative light in order to drive more sales directly to the OEM at higher margins. As explained earlier, the first indication that counterfeits are present is during a repair. If the OEM has both the supply and the repair contract, the end users will never be told that they have been using counterfeit equipment to avoid tremendous embarrassment. Yet if the OEM did not supply the equipment, it will gleefully announce even the slightest hint of counterfeits and attempt to use that information to reset the supply relationship back to its exclusive supply chain. The risk of equipment failure to the end user is identical, but only one set of pricing is competitive.

USED PARTS

All equipment is eventually used equipment. New parts are only available for as long as the OEM is still producing the product. Since many products remain in the field for years beyond the last date of manufacturing, users should realize that most parts they will have available will be used parts. The longer the warranty or postwarranty agreement, the higher the likelihood that used parts will be supplied.

Except for parts known to have wear and tear issues, such as actuator arms on disk drives or picker arms on printers, used electronics can be just as long-lived as new parts, with the caveat that the equipment pass all diagnostic tests and be inspected for leaking capacitors or other visible aging. It is well understood by the OEM and the independent service provider (ISP) community that pretesting parts before dispatching to the field for technician use is the best way to assure that the part will work on delivery. Parts that are not tested because they are brand new still need to be tested, as the incidence of "DOA" (dead on arrival) remains an issue.

OEMs seeking to increase sales of postwarranty or extended warranty service contracts may suggest that their service options will be superior due to providing only new parts. This is not likely to be true, even if the sales force believes it to be true. Once a product is no longer being manufactured, the source of truly new parts is a finite commodity. There is little financial advantage to holding onto a large inventory of spare parts. Most OEMs routinely restock their parts inventories for warranty support using repaired and tested parts. In cases of short supply of key parts, OEMs will purchase used equipment off the open market and scavenge parts if necessary.

OBSOLETE PARTS

With enough money and time, any part can be custom manufactured. There are several categories of parts that are in wide enough use that a volume run of parts is cost-effective. Many of the parts used in such common equipment as credit card swipe machines fall into this category. For most electronics repair purposes, small runs (anything under 1 million units is considered *small*) are simply impractical.

Due to the impracticality of manufacturing custom parts for lower volume uses, the vast majority of older model parts are repaired by specialists in board-level repair and returned to stock as spares. The companies that perform these services are rarely discussed but are a major resource for OEMs and ISPs for parts recovery. End users can also contract with these same repair specialists as their warehouse and supply of spares for end-user-sponsored self-repair programs.

For example, in the ticket kiosk industry, the items are in retail settings and subjected to hard use. Keeping this equipment up and running requires frequent adjustment and repair. Rather than have a unit under repair in a setting, the most common repair is a unit swap with the broken unit returned to a repair specialist. If the componentry is no longer in production, as is frequently the case, repairs are made by specialists at a detailed level to keep old models working. Experts in some industries report they can restore 95% of the parts for reuse in future repairs.

SUMMARY

Buyers have more options for acquisition of service parts than is generally understood. The OEM is one source among many, and the OEM is often buying service parts from the same sources as those they publically malign. Buyers seeking to keep assets in service beyond the warranty period offered by the OEM can combine nontraditional parts procurement with alternative labor sources to better control the useful life of their assets.

NOTES

1. iFixit (www.ifixit.com) specializes in providing consumers with parts and tools and guides for self-repair. The teardown photographs and parts lists were provided by iFixit with permission.
2. Google's purchase of Motorola Mobility in 2012 touched off a FRAND battle in both the United States and the European Union. For an excellent summary of the cases, see Jorge L. Contreras, "The Fraud Wars: Who's on First?" *Patently-O Patent Law Blog*, April 17, 2012, http://www.patentlyo.com/patent/2012/04/the-frand-wars-whos-on-first.html.
3. Comedian Steven Colbert did a superb piece on the absurdities of the case on his show *The Colbert Report*. See video at "Judge, Jury & Executioner—Copyright Law," *The Colbert Report* video, 5:14, November 26, 2012, http://www.colbertnation.com/the-colbert-report-videos/421501/november-26-2012/judge—jury—executioner—copyright-law. Had the case not been decided in favor of Kirtsaeng, vast swathes of the market for technology products would have been disrupted as the majority of parts, inclusive of copyrights, are manufactured overseas.
4. This section was quoted within the Defense Industrial Base Assessment of 2012, p. 105. The entire document is available at: http://www.theriac.org/pdfs/DEFENSE%20INDUSTRIAL%20BASE%20ASSESSMENT%20COUNTERFEIT%20ELECTRONICS.pdf.
5. See "Anti-Counterfeiting," NEMA, http://www.nema.org/Policy/Anti-Counterfeiting/Pages/default.aspx.
6. See "Anti-Counterfeit Policy: Revision 1—September 10, 2012," ASCID/NATD, http://www.ascdi.com/asna/vendors/counterfit_task_force/acpascdi.pdf.

10

Service Restoration
and Support Process

INTRODUCTION

This chapter elaborates on the sequence of events typical in a support or maintenance request. The role of each team involved in handling a service request is described, as are the different ways that the effectiveness of service is measured.

REPORTING

A failure of equipment or software must be reported in some way to begin the service restoration process. Most organizations have adopted some system to track incoming problems and make sure the calls are correctly routed for resolution. Often, but not always, these same systems are updated with the date and resolution of the problem, which can be a rich reporting resource for management.

Systems are also used by the service provider to track problem calls internally, and these systems do not commonly interface directly between the end user and the repair or support organization. Differences in how systems are used for inbound call tracking, call triage, remote diagnostics, and remote corrections to software are among the most important in how service delivery is accomplished between competitive options.

Figure 10.1 shows how the service restoration process progresses from the inbound trouble call.

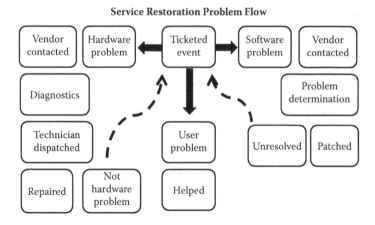

FIGURE 10.1
Service restoration process diagram.

CALL CENTERS AND TICKETING SYSTEMS

Most organizations have adopted a front-end call management system that serves as the first point of contact for anyone with a problem. The system may be as simple as a log kept on lined paper or on a spreadsheet, or more elaborate with calls routed automatically all over the world. The common functions are to document that there is a problem, then bring in the right resources to resolve the problem and to follow up until the problem is fixed or otherwise declared finished.

Once the name and contact information of the caller is input, and sometimes before a ticketing system is activated, the help desk/service desk personnel will try to resolve the problem without escalation. Assuming the problem needs more expert attention, the next step is to determine which team is the most appropriate to manage the problem. This is usually first differentiated by an evaluation of whether the problem appears to be a hardware problem or a software problem.

Despite manufacturers' assertions that their hardware cannot be repaired without a concurrent software maintenance contract, the first step in call triage betrays the fundamental difference between the two types of support services. The hardware problem will always involve a physical action: a set of "warm hands" must touch the equipment, whereas a software problem is never resolved by touching the machine. One of the main reasons that the help desk/service desk personnel will suggest a "power cycle" (turning off the machine and forcing a restart) is that if the power cycle restores service, then the problem cannot be hardware. The service request then proceeds down the decision tree intended to categorize if the problem is setting or software related, and then if software, which vendor is most likely to be responsible.

Once the call is logged by the help desk/service desk and transferred to the most likely support team, the provider of that service is in charge. The quality of the recorded information from the help desk is essential to efficient service delivery. Poor attention to detail and poor systems will cause service restoration to be delayed. For example, if a problem incorrectly calls for a technician to make a printer repair, but the underlying problem was a lack of the correct printer driver on the server, then none of the Service Level Agreements for efficient problem resolution will be met.

NONTICKETED EVENTS

Most systems are designed to assign a ticket or tracking number at the outset of any end-user contact. This discipline should not be skipped, as without a tracking capability, the system will be no better than a pencil-and-paper checklist.

Original equipment manufacturer (OEM) products equipped with a remote call-home system living outside the users' ticketing system are able to escape scrutiny of their equipment as to reliability and risk. Unless the end-user contracts for reporting on service events generated outside their tracking system, end users have no way to validate OEM claims of either service need (failure rate) or service response. It is possible for an event to be reported to the remote service office and the parts ordered and delivered days after the failure was invisibly reported, leading the customer to a false sense of security when in fact the redundant or spare parts were leaving the systems highly exposed to catastrophic failure for far longer periods than expected.

Specialist Hand Off

Once the specialist is assigned and the repair order with all of the relevant information is passed along, the service process begins. It is the contract holder's responsibility to locate parts and assign technicians. Some parts may be stored at the client location or within the field office, and others are held at specialty warehouses where fast delivery can be assured. This part of the service process is all logistics and process. The end user would be wise to understand how the service vendor has arranged its systems to support its service delivery promises.

Data hand off from a user to assigned service teams and downstream to the actual service provider is rarely standard, and as a result significant and useful information is often lost or buried in free-form text fields. Service tickets are often presented that require picking through paragraphs of "Bob says his printer doesn't work and I told him to reboot and that didn't work ..." combined with insufficient or erroneous information on the printer, such as the problem is reported with the wrong manufacturer name and an incomplete model number. Digging through this information to determine the basic requirements of a repair is a costly and unnecessary expense since summarization is simple.

The ticketing system should be used to refine problems and confirm models so that dispatch can be more efficient. The most frequent excuse I hear is that the end users refuse to provide the details when requested. This is a management problem and not a systems problem. Although service ticketing employees are probably low level, they cannot help the restoration process without collecting essential information. If the executive leadership team agrees that service response is important (which it always is), then everyone in the service delivery chain should be empowered to make sure that the service process is efficient.

Free-form text fields inhibit not only efficient service response but also thwarts using service management systems for management reporting. Free-form text is the enemy of reporting. In order for high-level business analytics to be run, both problems and resolutions must be consistently summarized so they can be scrutinized for trends. Although some text must be tolerated, text fields must be supplemented with summary fields that are limited in syntax and spelling.

SEVERITY AND CALL TRIAGE

It is common in large information technology (IT) organizations to allow for different levels of emergencies to be recorded so that the focus can be redirected to work on the most urgent and impactful problems before minor ones. The most severe problems are those where a critical system is down (often called "Mission Critical"), and the support agreements for these types of systems are generally written to plan ahead for such outages.

In a major outage, particularly when the cause is not clear, huge pressure is placed on managers to come up with workarounds, temporary fixes, and to restore service. In these pressure environments, vendors who are otherwise hostile to each other will be forced to work with each other. This is the situation where the powerful OEM is able to market its sole-source control of the entire system as a way to expedite return to service and to avoid finger-pointing.

There is legitimacy to the finger-pointing and multivendor lack of cooperation. However, there are multiple solutions to the fear of finger-pointing and that includes working with a multivendor specialist for repair (oftentimes not the OEM), as well as demanding the best from every vendor that has a footprint (or cyberprint) in the location. Demanding quality support from all providers is the role of good managers and can never be replaced with capitulation to vendor marketing.

The Service Level Agreement

Each service contract set (hardware and software) of the Service Level Agreement (SLA) lays out how problem notification is to be made (telephone, e-mail, weblogin, remote sensor), as well as how the user and the service provider are to pass along information. If the service process is inefficient, or missing key information, it will be difficult for either party to be satisfied with the outcome.

Many users only consider the pricing and response time elements of the SLA negotiation. In practice, it is far more important to have a process that runs smoothly rather than to strong arm a discounted price. It is imperative to ensure that the user provides excellent details about the assets and license to be supported, and not expect the service provider to guess. The more guesswork, the more opportunity for problems and the more likely the SLA will be unsatisfactory.

As in the example in Figure 10.2, if a trouble call is placed for an unspecified server or PC, someone has to track down the make and model before a technician can be dispatched. The time spent chasing details slows service response, and if questions are not resolved, the technician is far more likely to arrive without the correct parts. The more detail that can be captured in an asset database, the better the results for both the user and service provider.

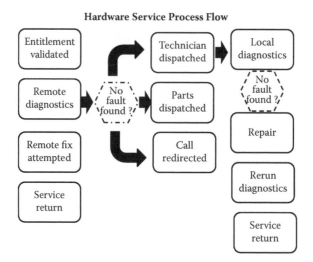

FIGURE 10.2
Data descriptions.

Not all detail is equally valuable. Too much detail also gets in the way if included in free-form text fields. If there is a choice between removing data and keeping data, err on the side of too much data. It is always possible to ignore descriptions of the color of the keyboard, but the brand of the graphics card cannot be reconstructed easily without looking into the machine.

Service Entitlement Database

It is unfortunately common for hardware maintenance contracts to be competitively bid, if they are bid at all, based on vague and inconsistent lists of hardware assets. The problem is that without detail, all bidders are guessing on the possible products it will need to support. There is no such thing as a *generic* PC any more than there is a generic car. We cannot get the slightest bit of useful pricing from an insurance company without disclosing the make, model, and age of the vehicle. The same is true for technology products.

For the most part, users are not intentionally sloppy about asset management. Errors creep into the asset or license database at all steps in the product lifecycle. Buyers fail to describe products fully to accounting, accounting fails to correctly book the assets, users add upgrades and features without informing the back office, service and support teams replace broken equipment with different models, machines are removed from service but not removed from the asset list, and parts are scavenged from machines in storage without mention.

However, if the user cannot identify what assets it has currently deployed, the management problem is quite large. In addition to the wasted costs of incorrect or inadequate service response based on poor information, there are other costs to the organization in addition to repair. Ghost assets are being depreciated, insured, licensed, and maintained. This is worth correcting long before writing new support and service contracts for hardware or software. For example, one of the largest banks in the world was able to trim 5% of its software license maintenance budget by carefully associating all maintenance renewals with a known asset.

Keeping asset databases updated is also thwarted by lack of resources, and not just the financial resources behind commitment to installing asset management software. Many older asset accounting systems were built when disk storage was precious, so the length of fields used for descriptions, configurations, and even serial numbers are often too short. Other systems, particularly modern asset management databases, are too flexible and allow the user unlimited descriptive fields. Neither product can function well as an asset management system without the naming and descriptive discipline being enforced by the end user.

The next most common reason for inadequate or sloppy asset tracking is a lack of syntactical standards on the part of the buyer. If a product description is not validated at the time of input, then just about any level of error can proliferate in asset databases. For example, in systems without a standard description for Hewlett-Packard as a vendor, the variations of spelling, abbreviations, and punctuation often create dozens of descriptions for the same brand name. Is HP the same as H-P? What about Compaq? DEC? (Both are manufacturers purchased by HP.)

Each level of description has its own variations. Once the manufacturer is consistently identified, the model number is often variously described. A single number, letter, or punctuation mark that does not visually register in the human mind is a completely separate model when sorted by a computer. Is the DL580 line the same as the DL 580 or DL-580? Until there is total consistency, none of the above descriptions would sort correctly for reporting.

Within the problem tracking system it is essential to carefully validate equipment descriptions, asset tag numbers, or serial numbers against a list of all covered (entitled) assets. A small error in description will delay support. For example, if a correlation is not made between a device, its associated licenses, and contract "entitlement," the service restoration process will be delayed as people scramble to search for missing or errant information. If the equipment description is inadequate, technicians will be dispatched without the correct parts or skills. The same is also true of incorrect software license and version level information. Searching for patches is highly dependent on both the operating platform and the full suite of installed software plus version level details.

Although some contracts may include an agreement to service the occasional noncovered device, this is highly inefficient and costly. Correctly connecting assets to specific contracts for both hardware and software is a wise use of time as each contract is added.

End users are always asked for asset records as part of setting up new service agreements and too often respond with a "This is the best we have" type of list. Users need to understand that the service provider must stock the right parts and have the right skills and certifications to complete a repair call. Failing to provide an accurate list is just as pointless as going to a Ford dealer demanding rush service on a Subaru. Asset lists exist in many places other than within IT. For example, the accounting department has records of all the original equipment purchases in order to correctly manage equipment depreciation, personal property taxes, and so forth. The accounts payable department has current invoices for all products under current agreements, and leasing companies have records of everything they show as being in service. It is already important for all these departments to periodically review such records and keep them accurate. If access to such information is blocked, then the problem is one of management leadership and not record keeping.

Admittedly, these records are challenging to regularize so that they can flow easily into a service contract bid or entitlement process. The more consistent the naming conventions used, the more easily the data can be handled. For example, it does not matter if a Hewlett-Packard DL-580 G5 is called out as a HP DL580g5 or Proliant DL 580(G5) so long as the same syntax is used. Wherever there is consistency, record descriptions can be manipulated electronically with easy find-and-replace searches and all three descriptions converged into a single description. Without standards, these three machine descriptions (Figure 10.3) will appear in three different places in any report.

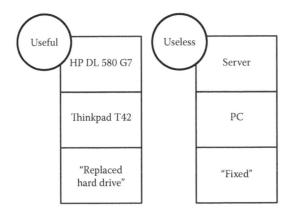

FIGURE 10.3
Ticketing descriptions.

DIAGNOSTICS

Diagnostics are an essential part of the service restoration process. In many cases, a hardware problem is reported to the end user from internal self-diagnostic routines, which then triggers the call to the service desk to begin the process. In less automated situations, the service desk may request the users run diagnostic software and report the results. Once a request for repair has been initiated, many OEMs have built remote diagnostics into their products so that the problem can be better identified, and specific parts and appropriately trained technicians can be dispatched. At the end of the repair cycle, diagnostics are rerun to confirm that the repair is complete.

Diagnostic software is most often delivered as part of the machine. There are some products, such as automobiles and aircraft, where diagnostic equipment is externally attached in order to read and report on service information collected from various sensors or self-diagnostics that pass information through to a central wiring harness. The manner in which the diagnostic routines are provided is less important than the function, which is to help the technicians to quickly and inexpensively return the item to service. Diagnostic software is as essential to repair as are service parts. If the diagnostic routines cannot be run or are unavailable for some contractual reason, no one can ascertain if the service parts (and the whole machine) has been properly restored. It may appear to run, but the standards set by the manufacturer are confirmed by the execution of the diagnostic routines provided for the purpose.

Contractually, diagnostics are usually provided as a form of machine code and rarely referenced in the original purchase agreement. If there are separate licenses for diagnostics, these most likely are associated with diagnostic tools and sold separately from the machine. Diagnostic code is sometimes provided on separate media, but most diagnostics are resident on the machine and executed when needed. Technicians and owners use the same access controls for running diagnostics as those for other forms such as applying microcode patches, restoration of firmware, or "reflashing" the basic internal operating system (BIOS).

In some industries, specialized tools are used in the diagnostic process. This is particularly common in the automotive industry where manufacturers have a long tradition of designing their own tools and interfaces. In the case of digital devices, this has created a situation where the independent technician is faced with the acquisition of multiple sets of equipment to perform roughly the same function on different brands. Some of this is being resolved by an industry effort to standardize interfaces but it remains an area of contention between OEMs and independents.

IT equipment has evolved with a more standard set of interfaces for data and product installation, often as a result of FRAND/RAND (fair, reasonable, and non-discriminatory/reasonable and non-discriminatory terms) codevelopment efforts. At this time, the self-diagnostic routines provided with the machine typically provide some indication of the culprit, such as a memory failure or an interface failure. Repairs made in the field are not so much diagnostic as parts replacements. If the replaced part does not resolve the problem, the next most likely part is replaced, and so forth, until a repair is successful. When parts do not readily resolve the problem, the most common resolution is to swap the entire unit. The possibly failed parts are returned to the depot for testing or repair.

More sophisticated specialty equipment is then used at the parts depot where failed parts are more completely evaluated and either returned to stock as spares, or the whole unit is repaired and then returned. The field technician is not expected to bring elaborate tools to the location.

If the machine is operational, the first step in identifying a product problem is to run diagnostic code. If the diagnostics run correctly, the triage of the problem will proceed to software problems, as broken hardware is very easy to diagnose. Assuming that diagnostics do not run or the report provides a specific flaw, the technician is dispatched with the replacement part.

Many OEMs have taken advantage of networking technology to allow the machine to self-diagnose and issue a phone call (or Internet contact) for repair without alerting the customer. This prevents the client from knowing about the product's lack of reliability, and also allows the OEM an almost unlimited time frame to arrange for a technician and associated parts to arrive without triggering the SLA response time. In some instances, OEMs have claimed that the remote diagnostics themselves are intellectual property (IP) making it impossible for owners to engage any form of service other than from the OEM. The legal status of diagnostic code as IP or patent remains unclear, along with other machine code functions.

The repair process, whether instigated by the end user or by a remote call-home service, eventually dispatches a technician and the associated part(s) or replacement spare so that the physical repair can be made. Often the replacement part needs to be refreshed with the correct level of code before it can be returned to service. At the conclusion of the repair, diagnostics are rerun to validate that the machine is operating correctly and that the unit is officially returned to service.

Access to all diagnostic code and the library of available patches is essential to the repair process regardless of who makes the repair. One has only to look at manufacturer Web sites such as Intel and Nvidia to see the scope of patching that is deemed essential by these manufacturers. OEMs that restrict access and claim copyright protection for diagnostics and related machine code patches are dictating to their clients that they can never repair their equipment without a continued service relationship with the OEM. The solution is for buyers to contractually demand in their purchase agreements that they control access to all diagnostic code without restriction as part of ownership.

SPECIALIST HANDLING

Once the service desk/help desk has assigned the problem to the applicable vendor, the technician provides the skilled labor for the repair and then has to arrange for the failed parts to be returned to the right party. It is not always clear which parts belong to whom, or who has responsibility for the removal, packing, shipping, and tracking of returned parts. For this reason, most small item repairs are managed with the assistance of a Logistics or Reverse-Logistics team.

Technicians

Hardware break–fix is a hands-on business. Labor is often the most expensive element of repair, so it is important to OEMs who provide warranty or postwarranty labor contracts to minimize the expense associated with labor. A technician dispatch is often more directly expensive than the parts (or machine) itself.

To reduce labor costs, manufacturers have been steadily designing products to reduce the skill level needed by technicians. There was a time when the field engineer (FE) was a highly skilled engineer equipped to diagnose problems, interface with product designers, and often jury-rig workarounds to return a machine to service. These guys (mostly guys) were often assigned exclusively to key customers and worked out of an office at the customer's location. For many years, the presence of an FE or CE (customer engineer) at the customer location was both necessary and a point of pride at being such a large IT organization that the OEM was investing in the customer.

Today's FEs are dispatched to the location with the failure already diagnosed. He then pulls and replaces the part, and leaves with the broken part to be returned to the OEM for autopsy. FEs do not diagnose problems; that is done with diagnostic routines provided by the OEM. They do not call the product engineers to resolve problems; they push and pull parts until the machine works. In difficult situations, they swap the whole unit and let the home office engineers sort things out later.

Problems that cannot be resolved readily in the field go through a series of escalations within the service organization. The majority of problems that require escalation are interface and interoperability problems between products. Managers are involved because fingers are being pointed as to which OEM is responsible for the problem and repair. These problems can be severe and result in an impasse requiring the end user to escalate matters within his organization to demand support.

Training and certifying technicians is variable by each OEM. The certification process for electronics is similar to that of automobiles. There are many sources of training, including self-training, trade schools, and, most important, on-the-job training. Technicians individually gather certifications as they work for various employers and are then hired by other OEMs and independents for their skills. Many OEMs offer training and certification to their authorized business partners as part of their requirement that partners have approved skill levels. Such skills are transferrable in much the same way that a mechanic in a Ford dealer can be hired by a GM dealer or an independent and be readily productive.

OEMs offering multivendor services are proliferating as customers seek to avoid becoming embroiled in OEM finger-pointing by hiring "one throat to choke." This has caused many OEMs to be faced with supporting equipment made by competitors. OEMs in this situation largely avoid paying their employees to become certified to support competitive equipment, which would enhance their resumes and make them more desirable to competitors, and instead hire independent service providers (ISPs) who will have hired and trained technicians for all the products in use.

OEMs have long been faced with supporting equipment outside of core locations. Some have taken the approach of limiting the geography of their sales to areas where they have resident technicians. This is clearly a problem for a growing company so many OEMs have determined they need to augment their geographical coverage with subcontractors. The result has been a steady expansion of independent technicians supporting a wide variety of equipment worldwide on behalf of almost all OEMs. Under the terms of these agreements, the technician is required to appear as an OEM employee complete with dress code, ID badges, and so forth.

This type of arrangement is increasingly part of a network of "flexible" labor where the ISP plays a crucial role in supporting the OEM as the feet on the street for the OEM. In all of the aforementioned hidden subcontracting relationships the ISP is prevented from announcing its role to the end user. OEMs are aware that users might take advantage of the obvious opportunity to eliminate the OEM layer of pricing and deal directly with the ISP.

Manufacturers do not want to compete for hardware break–fix because the margins involved are significantly higher than for equipment. In many cases, the hardware sale is a "Loss Leader" in that little margin is associated with the highly priced competitive equipment market, while service and support contracts remain far more lucrative. This is no different than many consumer-grade printers that are sold with ink that has more retail value than the printer.

End users do not have to participate in shifting profit margins within the OEM organization. Prices should be competitive across all markets, including the market for break–fix service. The primary responsibility for purchasing managers is to best serve their employers and not to serve the profit goals of the vendor.

Closing Tickets

If a problem is worth ticketing, it is also worth documenting as to the type of service provided and the parts used. Otherwise a major element of value made possible by the investment in the ticketing system is lost since little management reporting can be run off a list of problems that were "fixed." Problem ticketing systems are powerful resources that can be used to drive business analysis of many things:

- Analysis of the effectiveness of the help desk/service desk in doing its job
- Monitor the effectiveness of outside service providers doing their jobs
- Monitor the volume of service calls on similar products as a benchmark of product quality
- Develop metrics for product quality that will drive improvements
- Quantify need for service and support

For any of these metrics to be useful, the close out of tickets must be done with sufficient detail to feed later analysis. At the very least, the ticketing system should require a brief summary of the work that was done in a format that can be quickly sorted by product (to the model level), type of work (adjusted, aligned, replaced a part, swapped the whole unit), and which type of part (memory, hard drive, network interface card [NIC], motherboard, graphics card, etc.).

For software issues, similar categories should be used to summarize the root cause of the software problem ranging from user error to setting problems to version level conflict, and what was done, which includes tracking patches for each version of license, microcode and firmware, and so forth. This enormously valuable information is available and should not be discarded because the problem was "fixed."

When the user knows how often equipment of any particular model breaks in the field, the user then knows how much maintenance is needed for the product and can contract accordingly. Users will gain the critical information they need to buy the parts they need to stock for the most effective repair, and buy the labor contract that best matches the need. There is no other source of this information other than self-analysis and the repair closeout ticket is the critical link between what broke and what was fixed.

Improving repair close-out knowledge starts with improving how problem calls are organized. Problems must be categorized so they can be sorted. This applies to all problem reporting—for both hardware and software. Models or licenses must be consistently identified. All repairs or patches must correlate to a real machine or a real license, and not a generic partial description. It is a total waste of time to track problems without knowing which product had problems.

The specific model that was repaired must be identified because not all tickets begin with the correct product. The specific type of part that was replaced must be noted so that parts can be stocked accordingly. The part number is not particularly important. Part numbers are notoriously long, difficult to enter without making errors, and often subject to many variations depending on the source of the part. OEM part numbers are often different from distributor part numbers, which are undoubtedly different from compatible part numbers made by aftermarket providers. It is far more important from a management standpoint to know that 55% of the problems with Model XYZ are memory cards and 22% are network interface cards and 15% are power supplies than to be overwhelmed with part numbers without the ability to summarize.

For example, end-user managers may want to trap the failure rate of laptop disk drives as part of assigning a technology replacement cycle. The logic behind the idea is the reasoning that once hard drives start to fail in high volumes, it is a sign that the product is near the end of its useful life and should be refreshed. For this to be effective, reporting has to capture the category of a hard drive failure, not the specific part number. It is nearly impossible to calculate the incidence of hard drive failures within a given model of laptop if the technician is allowed to use any mutant description of their choice of spelling. "Hard drive," "HDD," "disk," and so on, are just a few of the many ways that a common problem can be indecipherable in service reporting.

Categories of Repair

Within the repair reporting summary, it is also useful to describe the types of action taken. Tracking actions allows managers to discern which products are more prone to certain types of repair needs. For example, a product with a high risk of data loss that is swapped instead of repaired on-site is a far more disruptive type of repair than dealing with a product with a less than optimal battery life. If given the choice, through analysis, most managers would agree they would prefer to own the product with the battery problem rather than constantly face worries about data loss.

Everything done by technicians in the field can be described using a verb. "Adjust," "align," "replace," "swap," and "upgrade" are all common descriptions. There are few ways that equipment can be handled leading to a short list of useful actions. This allows managers to develop metrics to evaluate service effectiveness as well as hardware quality by measuring the percentage of repairs made by replacing specific parts as opposed to swapping the entire machine. Printers with different percentages of adjustment needs would indicate different levels of quality.

Summaries of parts types are also easily resolved when consideration is given to the types of parts that might need repair for each specific type of device. It is not necessary to have thousands of potential failure types available for technicians to report; the only useful reporting are those part types that pertain to the specific type of equipment. For example, printers have rollers and trays and springs, none of which are parts common to servers or desktops. Limiting summary fields to the appropriate types of equipment also reduces errors and makes repair reporting accessible for management metrics.

MONITORING THE SERVICE LEVEL AGREEMENT

Most monitoring of the SLA is intended to make sure that the contract is met with respect to time frames promised. In the case of hardware repair, the SLA usually dictates the maximum time frame for a technician response within X hours before some form of penalty kicks in. For software, the response time to reach a software tech is usually measured in minutes. Time stamping has taken on more importance than the failure itself, which is a backward view of service quality.

The best SLA is one that is never needed because the product does not fail.

SLA monitoring by time stamp also obscures the issue of effectiveness. If the technician arrives within the SLA window but is unable to complete the repair in a timely fashion, the time stamp is irrelevant. Similarly, reaching a software tech that cannot resolve the problem and repeatedly transfers the call is a decidedly poor metric. The time stamp needs to be directed at the return to service, not the beginning of technician response.

Vendors are aware that they are being measured by a time stamp and not by effectiveness, and have taken advantage of this neglect by *gaming* their own agreements. Many agreements are based on the vendor agreeing that there is a problem, and not the user reporting the problem. The time stamp in this case can be delayed significantly (albeit innocently) by not agreeing that the problem is one that the vendor should handle. Other delays are introduced by the insistence that diagnostics be run before the problem is validated. These are reasonable requests on the part of the vendor but also part of the game.

Yet other vendors slide behind the SLA by using remote call-home functions that alert the vendor to a problem with a redundant part before the customer experiences an outage. The vendor then has time to ship a part and dispatch a technician without any fear that the SLA is being violated. This obviously works only with redundant parts that automatically fail over to a duplicate part in order to avoid an outage. The risk to the user is that they are never made aware of how much repair is needed and are at the same risk that they had with a single point of failure until the vendor completes the hidden repair.

Vendors slyly market this behind-the-scenes trickery as an advantage to end users, when end users should be made aware of all instances of product failure in their enterprise as a basic indication of service effectiveness. The vendor in these cases is taking hours or days longer than the contract specifies in exchange for premium prices and a delusion of excellent support.

There are users that have demanded, successfully, SLAs be based on measuring the time to repair rather than the time to arrive. This measurement is far more meaningful than the arrival time since any vendor can send a technician to stall for time. However, the more machines are manufactured to be redundant in order to mitigate the impact of part failure, the less need there is for a repair contract that is undoubtedly more costly than one with fewer real teeth.

Time to repair is a far more meaningful metric to software users but extremely difficult to demand. Developers of software are loathe to put their reputations or their billing on the line by guaranteeing problem resolution within any fixed interval. Having the commitment of the vendor to the "best possible effort" is probably as far as a contract can be taken, and this is admittedly wobbly language. The reasons are simple: not all problems can be fixed, not all problems can be made to recur (essential for diagnosis), and not every developer has the right people on staff at the right time to create fixes.

REVERSE LOGISTICS

After equipment has been returned to service, the final step in the repair process is "Reverse Logistics"—returning failed parts to the appropriate location. Reverse logistics is a messy process of packing, shipping, and inventory reconciliation. It is one of the areas where everyone complains. Users are not cooperative with returning equipment loaned to them under "spare in the air" agreements. OEMs are nervous that they are being abused by users taking advantage of warranty terms. Independents under contract to OEMs are under profit pressure to make sure as many parts as possible are returned for warranty credit.

Most OEM contracts require using an RMA (Return Merchandise Authorization), which allows the OEM to correlate the part return to the contract. Sometimes the technician may utilize on-site spares that need to be restocked. Parts are occasionally dispatched and returned to the local field office where they are then processed for repair or disposal. Very little is disposed at the end-user facility, as with the exception of physical damage, such as crushed equipment, most parts (over 90%) can be repaired and returned to stock.

Return Merchandise Authorization Monitoring

Behind the scenes of most repair organizations is the management of the return merchandise authorization (RMA) back to the parts provider or manufacturer. Many manufacturer warranties cover parts long after the parts and labor options have expired, and it is essential to get these parts returned and credits issued. Maintenance organizations have invested in systems to help assure credit for returns, yet this remains an area for improvement.

Individual users are deliberately motivated by a credit card charge to return products promptly, but larger organizations have to consider how their staff might better use their time for low-value assets. For example, it makes little sense to spend $25 in staff time to pull and package a $25 item, and less sense to invest in tracking systems and shipping expenses to do so. The best option for users in this situation is to focus attention on avoiding products that have high rates of return and include the costs of RMA handling in the evaluation of the total cost of ownership before investing in products known to need frequent replacement.

SUMMARY

Service ticketing systems are an underutilized management resource for evaluating service effectiveness for both hardware and software. Valuable and actionable insight into the basic reliability of devices and software can be gleaned from ticketing systems, so long as there is a commitment to data quality.

11

Building Blocks of the Machine: What Breaks and Why

INTRODUCTION

This chapter provides a background into the structure of common digital components and how they may become repair issues. Later sections of the chapter deal with repair issues in assembled products followed by a discussion of how common repair problems appear in particular industries.

BASIC HARDWARE COMPONENTS

Anyone who has looked inside a cell phone or personal computer will recognize the descriptions of the hardware components in this chapter. These same parts, at different scales, are used in products as varied as electric metering, digital radiology, movie theatre ticket kiosks, credit card machines, programmable coffeepots, and even exercise equipment.

All the products built using these components share common failure characteristics. A cursory understanding of the way the equipment works helps users to internalize the problems they face, or might face, when deploying equipment with electronic components.

Circuits

Circuits are pathways for electricity to follow, just as water follows a pipe from the street to the faucet. The most basic switch, like a common light switch, interrupts the circuit from ON to OFF or the faucet from OPEN to CLOSED. Computers are large arrays of switches of ON and

OFF positions controlling the flow of electricity to perform instructions. The word "binary" is used in computing because there are only two possible positions in a bit of information: ON or OFF. This remains the case regardless of the manufacturing techniques or scale.

The earliest computers were wired with copper wire—just as in a home—and the instructions were changed manually using a plug board as the programming. At the time, this was vastly superior to manual calculations for such things as artillery trajectories or code breaking the German Enigma code.[1]

Figure 11.1 shows the most common symbols used by engineers to design and describe the layout of circuits and logic for the manufacturing of integrated circuits (ICs).[2] Buyers of hardware and software have no need to pick apart circuit diagrams, but it is important to note that all machines are built using the same basic functions. The combination of logic gates (the operations on the right-hand side) make up the programming embedded in the chip, and the hardware parts on the left side support the flow of power to the chip.

The major innovation that has allowed the technology boom of the last 40-plus years was the invention of the IC and associated manufacturing techniques such as crystal growing and photolithography. The IC was a manufacturing leap replacing (integrating) soldering of wires (or plug connections) with circuits that could be printed en masse and combined to execute increasingly complex instructions at increasing speeds. ICs are the foundation of the entire information technology (IT) industry dating back to the late 1950s when the first patents were awarded.[3]

Common Circuit Design Symbols

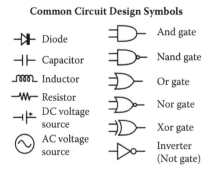

FIGURE 11.1
Common circuit design symbols.

Semiconductors

Semiconductors get their name from the types of elements and compounds used in the manufacturing of circuit boards known to have a range of partial (as in "semi") electrical conductivity properties. Elements with useful conductive properties include silicon, germanium, and carbon, and additional elements (boron, gallium, arsenic, phosphorus) are used in combinations to create compounds with valuable specific properties.

"Silicon Valley" got its name from the early semiconductor manufacturers clustered in the San Francisco Bay area that were using silicon crystals as the most popular basic framework for building ICs and assembling them onto printed circuit boards (PCBs). Although much semiconductor manufacturing is now done overseas, and many elements other than silicon are used, the process is the same as it was decades earlier.

Simple Semiconductors: Transistors, Resistors, and Diodes

At the most basic level, semiconductors are simple devices manufactured to provide a single function, such as a diode in a battery operated flashlight controlling the flow of current when the bumps at the ends of the batteries are correctly aligned. (Diodes control the flow of electricity in only one direction.) Transistors are multiple diodes that are able to control the flow of electricity in more complex routes much like highway interchanges are set up to control the flow of traffic in a predetermined fashion. Moving up the complexity scale, combinations of transistors are designed to operate as "logic gates," which are the fundamental functions of a computer. These parts can be built individually and soldered together, as in early computing, or manufactured as ICs.

Integrated Circuits

The innovation in manufacturing that became the integrated circuit (IC) was the result of using advanced crystal growing techniques with photolithography to "print" logic gates into a semiconducting silicon base. The crystals are not the logic gates of the computer but are thinly sliced to be the substrate for printing of what had been wired transistors in miniature. Printing logic gates are vastly more efficient than soldering connections. Using thin slices (wafers) of pure silicon crystal (or similar) combined with other conducting compounds in a process called "doping" took advantage of the opportunity to layer more gates into the product.

The ability to build complex circuits using photolithography rather than manufacturing technology is the key to both low unit cost and miniaturization. The most technically sensitive of the processes is growing the crystals to a very high quality since flaws in the crystal would allow errors in printing, which would destroy the consistency and integrity of the chip. As long as ICs are manufactured using high-quality crystals and the same photolithography, all ICs from the batch will theoretically be identical. Actual practice shows that not all manufacturing is perfect and that there is a requirement for quality control.

The manufacturing plant and equipment that are needed to manufacture ICs and chips are very expensive. These investments are continually retooled to keep up with the market, replacing both the size of the crystals being grown and the associated production processes. Constant competition and market demand have been following the track of Moore's law, whereby the capacity and capabilities of the products are doubling every 2 years. Moore's law may eventually hit a threshold of the limitations of silicon, but other materials and particularly "nanotechnology" are expected to expand the limits of what can be done with ICs.

Chips

Integrated circuits are built to function as memory, controllers, graphics processors, and, of course, microprocessors. Chips are combinations of ICs designed for a specific function, such as a processor, and manufactured as a monolithic product. The chip itself cannot be subdivided. Each chip is intended to be installed as a unit and is typically presented with wiring that allows the chip to be installed as a unit into a PCB, more on this is provided in Chapter 12.

The pace of innovation in the chip and IC industry is so fast that these leaps in capacity quickly obsolete products that were the state of the art just a few months before. Once a new wafer size is in production, the chipset design can be more elaborate, and the production capacity of the manufacturer will be quickly devoted to the newest and most desirable products. As producers upgrade their factories to keep up with competition, many facilities cease being able to produce products in the now "obsolete" format.

It is rarely practical to reproduce older model chips once the lines are retooled, so all parts and spares that are ever needed will need to be ordered during the production period. For some low-volume products, the entire production run may be a single pass. This has obvious impacts where ICs are intended for long-term use and not throwaway consumer products. Appliance manufacturers, for example, have to deal with the reliability and durability of ICs that can no longer be bought from the original manufacturer.

Capacitors

Capacitors are miniature batteries and as such essential to electronics and computing because of their ability to hold a charge. A wide variety of capacitors are installed on almost every PCB for every size imaginable. Capacitors provide the "muscle power" to manage variations in power and protect components from spikes and dips in power. Their ability to quickly release electricity in a burst powers camera flashes and small motors. They are deployed as tiny batteries to supply a consistent charge for such elements as volatile storage. They are critical elements in power supplies and uninterruptible power supply (UPS) systems.

Capacitors are cousins of the conventional liquid-filled automobile battery. Liquids known as electrolytes are separated by a thin film that controls the flow of electrons between liquids. As capacitors age, the electrolyte eventually evaporates. Leaking capacitors lose function and can damage the PCB itself.[4] Just as the car battery loses cranking power in cold weather, capacitors also suffer from degraded performance in extreme cold. Overheated capacitors can burst and catch fire.[5] Variations of voltage, as with poor electrical quality, degrade capacitors more quickly.

Capacitor failures are avoided, where possible, by specifying capacitors with higher ratings than required. More robust specifications add cost to the unit, so the higher reliability product will probably carry a higher unit cost than one using the most minimal options available. Failure rates of capacitors are variable based on age (fresh is best), ambient temperature (consistent room temperature is best), and variations in voltage (swings in voltages reduce capacitor life). Items rated at 350,000 hours (roughly 40 years mean time between failure [MTBF]) are degraded by half with a 10 degree change in case temperature (20 years MTBF).

Printed Circuit Boards

The printed circuit board (PCB) is the skeleton upon which ICs, chips, and other parts are combined into a functional device. PCBs are used as a substrate for connections because the manufacturing process facilitates wiring of thousands of miniaturized connections. PCBs replace what would be very large and unwieldy wires and soldering points into a simple board, which is ideal within a wide variety of products.

Manufacturing PCBs is based on using a combination of metal foils (mostly copper) and insulating layers coated with epoxy resins, which are then coated with a solder mask (usually green) and then printed, etched, and otherwise processed to prepare the board for assembly. Specified components are then added to the board, and the resulting product coated to protect the connection points.[6] Because the PCB is manufactured by a printing technique, the U.S. Copyright Code has been involved in protecting the interests of the printer (the manufacturer) by providing special protection for the masks used in printing. This has not been particularly effective to help the industry prevent patent infringement, which has been better protected by the enormous capital commitment than any special protections under copyright law.

PCBs can and do fail due not just to static charge (the reason for antistatic packaging), but also due to problems of design, materials degradation, frequency of setting and resetting components, and even vibration.[7] Problems of aging are particularly difficult to catch once in the field, particularly since the vast majority of products are not tracked for failure beyond the initial warranty.

The supply chain also aggravates difficulties with determining flaws of design, component quality, or manufacturing error. Most PCB manufacturing and assembly is currently done in Asia by specialist companies. Original equipment manufacturers (OEMs) from all over the world contract for manufacturing or assembly at the same facility. Many components are bought in bulk according to specifications and vendors for components may vary by batch. Once problems crop up in the field, the production run has long ago been completed and it may be impossible to discern if any remaining components are still in the supply chain.

Board-level failure (including IC, chip, and capacitor failure) is an easy repair provided that a replacement board is readily available. Little, if any, board-level repairs are performed in the field. Most repairs are made by swapping the board and then returning it to the OEM under warranty.

For products out of warranty the owner can contract with a specialist in board-level repair to return the failed item to inventory as a future spare. OEMs use board repair specialists for the same reasons as end users, because once the production run is completed for the chips or boards, no new parts are available.

As boards have become cheaper, and as more functions have been built into boards, very few products are preemptively modified using engineering changes (ECs). Most problems with boards are not defects of the "recall" variety. Board failures are usually dealt with by the OEM as part of the normal support contract, and little or no effort is made to alert the user that a consistent problem exists. Since tracking of electronic reliability at the field level is currently absent, OEMs are able to easily hide issues of component failure.

Power Supplies and Fans

The most common failures[8] across all machines are the power supplies and fans. Power supplies are such common points of failure that development of redundant products initially focused on power supplies ahead of other parts. Even when it was tremendously costly, redundant power supplies and hot-swappable power supplies were widely adopted.

The causes of power supply failures are well understood. Power supplies include some of the most vulnerable of components: capacitors and transformers. Each is known to degrade rapidly under conditions of heat and voltage fluctuations, both of which are common in all but the most carefully controlled settings. Many power supplies are cooled by air, linking fan performance to power supplies.

Fans are known to have short lives as mechanical devices, adding to the vulnerability of the power supply to the system. The action of the fan is mechanical and subject to wear, so fans have very high failure rates relative to electronics. Fans also clog easily with dust and debris, which can easily interfere with motion and effectiveness. There are very few parts in a computer where "Preventative Maintenance (PM)" is legitimate and fans are at the top of the list. In this regard, PM for fans is simple housekeeping.

Alternatives to air cooling with fans exist. Water cooling is a good viable alternative, but has similar mechanical action and wear problems. Pumps and control valves are areas to watch for failure and service issues including leaks and corrosion.

Batteries

Many products now include some form of internal battery to carry settings between periods of use and power supply drops. Batteries are a common point of failure, as they often lose charges more rapidly than planned, particularly when the drain is high, as in cold temperatures. It is best to replace batteries before they fail, but testing battery strength and life to determine lifespan is just as time consuming as replacing the battery itself. Most choose to replace batteries on a schedule.

Batteries and their tiny cousins, the electrolytic capacitor, are prone to leaking as the result of either aging, poor quality, loss of seal, and physical damage in handling. Leaking is a serious problem that can destroy connections within the PCB and potentially cause fires. Outgassing, which is also a result of tiny leaks, is more common over age and eventually degrades the battery much like canned sodas that eventually go flat when stored.

Controllers

Specialized chips are produced to manage the interaction of peripheral devices and the main system. The functions used to be performed by externally attached machines called "controllers," but as equipment was miniaturized the controller functionality eventually became small enough to be housed on the same PCB. In simple machines, controller functions may be integrated within the main processor.

The function of a controller is to manage the I/O (input/output) between the main processor (CPU) and the peripheral. This can include communications (a network interface card, or NIC), disk storage, tape drives, printers, and even controllers for display devices such as liquid crystal displays (LCDs). Most controllers are now either separate PCB "cards," or onboard controllers built into the chipset. Onboard controllers are less favored for higher capacity applications as greater functionality dictates more capacity available through a separate card. The reasoning is simple. Each I/O request uses CPU cycles. The more cycles devoted to functions that can be performed externally, the fewer the cycles available for processing by the CPU.

It is for this reason that high-intensity graphics applications benefit from taking advantage of a separate graphics "card" (really a graphics coprocessor) rather than an "onboard" graphics function designed into the chipset. The ability of a machine to separate functions that can be run concurrently greatly enhances the overall performance of the device.

Controllers often include some form of buffer or cache memory to operate at separate speeds. In a typical I/O request from the processor, the controller initiates the request to the storage device, stores the returned information in some form of memory (cache or buffer), and then returns the requested information to the processor at the speed available through the channel. This allows data retrieval to happen concurrently while different peripherals operate at different speeds.

External Connections

It is hard to imagine today that most original computing did not have any external communications interfaces. The products that attached to the processor were all directly (locally) attached. Obviously large changes took place in the ability of users to connect to computers from increasingly remote locations, such as locally cabled terminals and printers, all of which required the use of front-end processors to collect information and requests for processes from remote locations and translate them into centrally executed functions.

All of the products used in today's vastly interconnected cyberspace still use the same components as in other digital products just in different patterns. The quality of the cabling and connections has an impact on machine performance.

Central Processing Units

The central processing unit (CPU) has changed little in basic organization since is origins in the 1960s. Every CPU is a combination of memory (where instructions are stored for fast execution), registers (where instructions are processed), and I/O (the retrieval of instructions or data, and return of results back to some form of storage). Each instruction is an action, and processors are rated for speed based on how many instructions per second they can perform. The faster the CPU, the more instructions can be processed in a second. The more CPUs that can be ganged together, the more can be done at the same time. Multiple CPUs can be assembled together onto a single chip, and those chips also used in combinations. Raw computing power has followed Moore's law for several decades due to these advances.

CPUs are extremely sensitive to heat and breakdown quickly if not cooled. Older mainframe computers generated so much heat that water cooling was the only effective way to keep them running. (Water has a higher thermal transfer rate than air.) Water cooling is coming back into vogue because of its efficiency in high-density computing, although in smaller formats usually without placing chillers on the roofs of buildings. Air-cooled processors contain fans to keep air flow circulating and a fan failure can quickly "fry" a processor. Because CPUs have no moving parts, failures are rare without cooling or voltage issues.

The more CPUs packed on a board, the more vulnerable the assembly is to heat. Whereas in a distributed server environment a single processor running a single application might fail and take down a single application, a card carrying multiple CPUs when overheated is far more likely to take down the whole card full of processors and impact not just a single application. The pendulum still swings between high-density cards and high vulnerability.

Specialized Processors

Many machines are built with specialized CPUs deployed for support tasks, such as graphics, cryptographic, and arithmetic purposes. These CPUs are still CPUs but are developed to perform particular tasks. For the purposes of support and repair, they are indistinguishable from any other CPU.

Memory

Memory is another form of an IC that is intended for use within a processor for short-term storage of either program instructions or data. The physical capacity of memory determines how many instructions can be immediately available, which determines the speed at which programs can execute.

As an IC with no moving parts, memory is highly stable and rarely fails without some outside influence, such as excessive heat, voltage swings, or vibration, which might loosen the connections to the PCB.

Channels

Movement of externally stored data and instructions flows through a channel into the CPU. The channel is a portal with a fixed capacity, much like a water pipe in a home. A small diameter pipe does not carry as much water at the same time as a larger pipe. The same is true for channels.

Developments in processor speed that are not matched with developments in channels result in poorly balanced machines that tend to "clog" waiting on operations on one or other side to complete.

RELIABILITY ISSUES OF COMPONENTS AND MATERIALS

Unless subjected to excesses of heat, moisture, or variations in voltage, most ICs are very stable and have very low failure rates. Their stability is a happy consequence of the manufacturing process and not because high reliability is itself the goal.[9]

Problems arise immediately once the chip meets the real world where heat, moisture, and voltage are constant variables. Voltage control devices, such as power-conditioning equipment and surge protectors, are widely deployed to protect sensitive circuitry. Moisture, such as excessive humidity, can condense on circuits and create connections where none were intended. Data centers are climate controlled not for the comfort of the personnel, but to keep the equipment at both specified operating temperatures and also at relatively low humidity.

Most failure issues with ICs and chips are related to the need to provide consistent ("clean") power to the circuit and remove excess heat. *Clean* power is essential to removing variations in voltage that are damaging to semiconductors at both the low and high end. Ordinary household current supplied by the local electric utility company is far more variable than most consumers understand. For this reason, most data centers routinely install power-conditioning equipment as well as battery-backup power systems and standby generators. Consumer products are often powered through external transformers and surge protectors.

Heat is a constant battle, as the higher the temperature, the slower the circuit. Excessive heat can permanently damage circuits, commonly called "frying." Many ICs are cooled using fans where air cooling is sufficient. More effective heat transfer is provided by circulating chilled water within the case, an old technique from the 1960s mainframe era now coming back into vogue with the extreme density of boards now in use. Gaming enthusiasts routinely modify their computers to control heat and in so doing push the envelope of what can be done with ICs when conditions are optimal.[10]

Component quality is variable. Differences in both component specifications and manufacturing quality will cause different failure rates and types of failures. The pace of innovation in manufacturing techniques and circuit design will continue to quickly obsolete products with implications for availability of parts, support, and patching.

There are reliability and durability issues for every component, including the substrate used for the PCB. Not all plastics and resins have stood the test of time. Not all coatings intended to protect components from moisture are equally effective, and not all components are installed correctly. A slightly insecure connection can easily work loose during ordinary use, taking down the component intermittently and eventually permanently. Cables and connectors commonly work loose and are often implicated in repair calls.

SUMMARY

Digital machines are mostly combinations and variations of the parts described in this chapter. Although there will be continued improvements to the speed of processors, density of storage and media, and new methods of manufacturing, the essential elements of digital designs remain the same. The more buyers understand about the products being selected, the less they can be misled or confused by vendor marketing claims.

NOTES

1. The word "computer" came from the job description of the women hired to calculate the trajectory of artillery shells during World War II. They were making repetitive calculations/computations and were known as Computers. When these functions were automated using the earliest electronics, the word was readily associated with their function.
2. The symbols in Figure 11.1 are widely available in computer-aided design (CAD) programs for use in design. This particular image appears in Wikipedia and many other Web sites without attribution. The source of the copyright of this image is unknown, but the list itself is not proprietary.
3. For a brief history of the invention and patenting of the first ICs, see Mary Bellis, "The History of the Integrated Circuit aka Microchip: Jack Kilby and Robert Noyce," About.com, http://inventors.about.com/od/istartinventions/a/intergrated_circuit.htm.

4. See American Power Conversion (APC) White Paper on avoiding capacitor failures. American Power Conversion, "Avoiding AC Capacitor Failures in Large UPS Systems," White Paper #60, 2003, http://www.apcdistributors.com/white-papers/Power/WP-60%20Avoiding%20AC%20Capacitor%20Failures%20in%20Large%20Systems.pdf.

5. Dell was embroiled in a $200 million lawsuit stemming from the production of over 11 million desktops with a known problem with capacitor fires. The origins of the capacitors implicated in the problem have been traced back to the mid-1990s. The capacitors at issue are used in many products and the full impact of the particular part is not yet known, although as time goes on, the inventory of old parts is most likely consumed and will not specifically appear in more recent products.

6. *Wikipedia* offers a well-written and accessible article on the PCB manufacturing process. See "Printed Circuit Board," *Wikipedia*, http://en.wikipedia.org/wiki/Printed_circuit_board.

7. The Reliability Information Analysis Center of the Department of Defense openly discusses PCB MTBF at http://forum.theriac.org/showthread.php?p=31843.

8. TekTrakker® failure rates confirm anecdotal evidence that power supplies and fans are the most frequent items to fail in common computing products.

9. Reliability is not of particular value when ICs are included in throwaway devices such as disposable cell phones. If a low-quality chip could be produced at a lower cost, the industry would quickly adopt it.

10. Gamers have neatly explained the physics at a level I do not care to delve into. See "Info: Why Does Too Much Heat/Voltage Damage the CPU—Scientific Version," Overclock.net, http://www.overclock.net/faqs/19390-info-why-does-too-much-heat.html.

12

Repair Issues by Product Type

INTRODUCTION

This chapter deals with the ways machines fail due to the commonality of parts and use of contract assembly facilities, primarily in Asia. Because most digital electronics are variations on a theme, as discussed in Chapter 11, failure and repair also falls along common lines through hundreds or even thousands of different products.

PRINTED CIRCUIT BOARDS

The printed circuit board (PCB) is the substrate upon which everything else rests. The PCB provides connections between components using traces (looks like tiny wires) and vias (holes). The PCB also insulates components from heat and accidental cross-connections. Because the PCB is itself a manufactured product, all of the potential flaws of manufacturing apply to PCBs as well as more complex parts.

PCBs combine many materials any one of which can be of lesser quality than expected. Different types of solder have different properties, some of which impact the flow of electricity and some are so brittle the joint can crack apart and fail. Rough handling of PCBs both during and after manufacturing can cause connections to fail. Different plastics and resins used in the board have different properties that determine if the board can withstand heat without pulling apart (delaminating). Extremes of heat, moisture, and voltage can cause chemical reactions between the component materials, which can cause failure.[1]

CENTRAL PROCESSORS

At the heart of all digital electronics applications is the central processor unit (CPU). All CPUs have similar failure characteristics because they are essentially the same product produced in different sizes. Multiprocessors are literally multiple CPUs installed on the same board. Faster processors are just that—faster. Each iteration of new chip design requires changes in the manufacturing tools for producing integrated circuits (ICs), but the concept is still the same. New production techniques will change the ways that CPUs fail, but failure there will be.

Currently known problems with chips and failure fall into three major categories: environmental damage (heat, moisture, and voltage), manufacturing defects, and machine code defects. The user is only able to control the environment. The other two problems are completely under the control of the manufacturer.

Environmental Damage

Chips are particularly vulnerable to extremes of heat, which can easily cause complete and permanent damage.[2] Depending on the packaging of the chip (not the retail packaging for purchase), the chip may be more or less vulnerable to extremes of heat damage. The lower the melting point of various plastics used as casings for chips, the more likely heat will melt the parts surrounding the chip and destroy essential connections on the PCB. There are other very technical reasons, such as tunneling electrons, that impact chip performance and damage due to heat that are outside the scope of this discussion. Suffice it to say: Heat is bad.

Voltage variations are a problem for chips because of the heat generated by surges of power. Low voltage conditions are not damaging but can cause equipment to stop working because there is not enough power to keep the equipment running.

Moisture is a problem for electronics because water conducts electricity extremely well. The same safety issues that necessitate warnings on hair dryers to avoid use while in the bathtub apply to everything electric. Even without a risk of electric shock, moisture on a circuit causes "shorts" in the circuits, leading to unexpected results. For example, my refrigerator was equipped with a tiny circuit that opened the water dispenser when a button was pushed. One summer, I was on vacation and turned off my air conditioner while away. When I returned, the high humidity of the weather had caused the circuit to close without anyone around. The result was a flooded kitchen.

Water damage is obviously a major issue for users of mobile cell phones that are equipped with moisture sensors to prevent users from making warranty claims for units that have been thrown into swimming pools, dropped into toilets, or had beverage damage. The same problem occurs for all electronics, not just consumer products.

Manufacturing Defects

Manufacturing of chips is a high-precision industry but not a perfect industry. Silicon crystals are grown to be extremely pure but less than perfect crystals are still used. Tiny imperfections in crystals carried into chip manufacturing can create unpredictable errors downstream. Not all companies operate with the strictest controls nor are all companies equally diligent about quality control. Some might not win any awards for high ethics. Manufacturing defects are possible from every step in production.

One of the reasons that end users should avoid using consecutive serialized products in redundant settings is the possibility that the same error on one serial number will have been repeated in the same manufacturing batch. The closer in sequence each serial number is within the batch, the more likely the same errors can be introduced into the product at the same step. This was driven home in a devastating outage experienced by RIM (now BlackBerry) when two network servers went down for the same problem within a nearly concurrent period of time. The two servers had been deployed at the same time, built at the same time, and were in use simultaneously for the same period of time as backups to each other. Unfortunately, they suffered from the same latent manufacturing defect and failed at the same time.

Machine Code

Most chips are preprogrammed to perform their function by a combination of printed logic instructions and the use of variable instructions known as "Machine Code" or synonyms including "Firmware,"

"Microcode," "BIOS," "Programmable Logic Code (PLC)," and "Internal or Integrated Operating System (IOS)."

Machine code is used to control the most basic of instructions within the machine, the sequence of start-up, the transfer of information between locations, and the execution of mathematical chores, all of which are unique to the machine to make it suitable for use. One does not *shop* for or *select* machine code; the code comes with the machine and must be functional before anything else can be added.

The variable and programmable aspects of microcode, if there are any on a machine, are used to allow for defect support, setting controls, and interface adjustments. These actions could be set into the wiring of the machine, but if there were modifications required, then the part itself would need to be replaced in the field. The most common type of defect corrected by the original equipment manufacturer (OEM) to deployed chips is done by updating the machine code for the product. For more on machine code, see Chapter 8.

Because of the peculiar relationship of machine code to both the hardware chip and the software controlling the use of the chip, the buyer needs to take time to dig into the policies of any proposed agreements and make sure their particular requirements for support are met. The most practical approach is to require that defect support associated with machine code is treated as hardware. This allows the owner of the equipment to behave like an owner and contract for services as they see fit, sell or buy used products as they see fit. If the OEM claims that all machine code (microcode) is intellectual property (IP), whether true or not, the owner of the equipment is really only a licensor of the hardware, as support can never be severed from the OEM, nor can any such items be sold as used equipment without the permission of the OEM.

PERIPHERALS

Equipment attached externally to the CPU is known as a peripheral. Some of the functions that used to be managed as peripherals have been adapted to be packaged inside the same cabinet as the CPU, but they are still peripherals. Common peripherals are tape drives, disk storage, networking equipment, and print or other output devices (Figure 12.1).

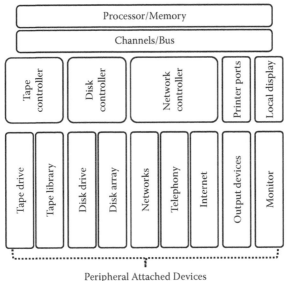

FIGURE 12.1
Peripheral attached devices.

Tape Drives

Tape drives have been a mainstay of long-term data storage for decades. Although tape has evolved from a common storage format for daily batch processing to largely archival storage, tape remains a staple backup medium. Storage and retrieval of data sets stored on tape has been semi-automated with the advent of robotic tape libraries. Robots have taken the place of human operators who used to shuttle between tape storage vaults and individual tape drives. The management principles remain the same.

When a backup is scheduled, a "scratch" tape is identified and mounted in a tape drive. Data is then dumped onto the tape media, and the tape returned to a position in storage (or ejected for external storage). This remains a very physical process with a high rate of failure, which is difficult to avoid. Media can jam in the tape drive. The tape drive itself can jam or fail (there are PCBs with capacitors in these too). The robotic arm can jam, the track on which the robot moves can be displaced, and the spaces that the tape media itself is stored can be misaligned. Tape media itself can fail in rather ordinary ways such as stretching, delaminating, or breaking.

The use of tape as backup media is on the decline, as immediately available disk storage decreases in cost and the opportunity to transmit archival data sets to spinning storage for backup improves. The demise of tape drives has been predicted for decades, but tape remains viable because of the tremendous cost-effectiveness and simplicity of shipping tapes to archival storage.

Failures and repairs of tape drives are rarely mission critical for two reasons. First, the backup process is a scheduled event that can be rescheduled to a different time. As long as the data set(s) are to be archived, the date of the dump to the archive is not always essential. Many robotic systems use an intermediate disk storage array device to aggregate tape data sets before writing to tape. This is itself a backup media.

Second, most tape drives are purchased in multiples so that if one drive is down for repair, another is still available for a backup or restore. Operationally, most organizations expect a high-failure rate for tape drives and plan ahead for one or more drives to be unavailable. Less hardware failure would do much to reduce the volume of equipment needed in standby mode.

Disk Drives

Disks are named for the shape of the platters used in the original disk devices. Drives looked much like a stack of old-fashioned "records" (think classic vinyl) with small spaces in-between for tiny mechanical arms, called actuators, to write or retrieve data from the magnetic coating on the platters. The platters spin and the speed of the rotation is a major factor in the rate at which data can be retrieved or written. Thus, a 7200 RPM drive is slower than a 10,000 RPM drive. The other major factor in seek time is the location of the actuator relative to the location of the data. Larger form factors are inherently slower if the actuator arm has to move across a larger surface.

Problems with disk drives fall into just a few categories, most of which we know as a "head crash." First, and least catastrophic, is a failure of some of the sectors of the media. Bad sectors, unless occurring at the location of the basic internal operating system (BIOS) and start-up instructions, do not prevent the disk from being used, but they are not normal and indicate that the disk needs replacement. A failure of either the actuator or the rotation of the drive is what causes a *head crash*. Regardless of the cause of the head crash, the data on the disk is rarely retrievable and the equipment is decidedly broken. Most data loss is a result of a head crash.

The same mechanical problems with hard drives apply regardless of the diameter of the platter or speed of rotation. Prior to the disk array concept, advances in media density and rotational speed allowed steady growth of the format of enterprise-scale disk drives to the point where customers began to be concerned that a single head crash could seriously impact their enterprise. Users adopted a variety of techniques to buffer themselves from catastrophic data loss on disk drives ranging from constant data backups to tape, offsite disaster recovery programs, and even mirrored data sets. Many users were wisely worried about *putting too many eggs in one basket* (a recurring theme).

The combination of concerns about too high a density of the disk drives and the availability of the miniature PC version of a disk drive came together in the RAID disk array products now in common use. These products aggregated multiple inexpensive disks in a format where one, or more, of the disks serves as a spinning backup for the others. The early versions of RAID (literally stands for Redundant Array of Inexpensive Disks) were crude where each disk drive had a complete backup. This was known as "mirroring" and became known as "RAID 1." Eventually, more sophisticated algorithms were developed to reduce the amount of physical additional capacity needed to effectively buffer users from the potential data loss as these many "inexpensive" disks failed more frequently than their exquisitely engineered enterprise cousins.

It is important to note from a physical reliability and failure rate standpoint that the disks are not themselves highly reliable. They do not have to be as long as the service team continually replaces the primary failed disk before the backup disk itself suffers a failure. Much more work has been put into the data management side of these products than the actual drives. The result has been a product set that is not notably reliable, but has mostly buffered end users from catastrophic data loss. "Mostly" is a deliberate term, because with any best laid plans there are always situations where a combination of unforeseen events align to cause data loss.

Users of such products need to know that the maintenance process begins with the equipment itself. Most arrays are equipped with "call-home" features where the equipment itself issues a trouble call to the vendor. Unless programmed otherwise, the owner (user) does not know that there has been a part failure and the failure event is not managed by the user's service ticketing system. This allows the vendor to ship a part from a part depot to a field engineer who can then arrive with the correct part in hand to replace the failed device. Since the user did not know of the trouble call, it is exceedingly difficult for the end user to know the actual failure rate of the equipment under these circumstances.

None of this means that the equipment is lousy or that the vendor is inattentive. What it means is the end user has tremendous difficulty in extricating themselves from vendor supplied service contracts. This reliance on OEM support has downstream implications for the secondary market and preservation of residual value (or any value) for the owner. In cases where the OEM is the sole provider of support, the OEM can easily create policies and fee structures that severely restrict any attempts at resale.

Solid State Disks

Despite being named a disk, solid state disk (SSD) products are not spinning platters of magnetic media. The advantage of an SSD is the absence of the major points of failure with spinning media—the actuators. Although initially expected to be a panacea for ending head crashes, the SSD is not the perfect substitute for spinning media due to limitations of design (the issue is one of "write endurance"), which may or may not be overcome in the near future.

SSDs are flash memory chips made to appear to the operating system as a spinning hard drive. The packaging mimics the shape and attachments of a spinning disk, which allows for an immediate substitution. SSDs are by design more reliable than their spinning counterparts. When appropriately configured, they can also improve seek times (since there is no latency for actuator movement) and data transfer rates. The SSD can also be built from nonvolatile storage making the device more independent of power loss and less demanding of backup and recovery protocols.

Repair of the SSD is similar to repair of any other type of memory. The bad stick is removed and thrown away and a replacement provided. Backups of data are obviously important, just as they would be for conventional spinning disks.

OUTPUT DEVICES

The most common output device is a printer. Print and related paper handling equipment remains the most highly mechanical environment in computing. Printers are now a converged technology with copiers, scanners, and faxes. Depending on which type of organization developed the product, printers will be priced and serviced as printers with on-call repair models or as copiers with repair contracts based on usage.

Regardless of the type of print head (impact, inkjet, laser, thermal, or xerographic), the vast majority of printer problems are related to paper handling. Paper is full of dust and dust jams mechanisms. High-volume printing is often relegated to special rooms with advanced dust control systems to avoid spoiling other electronics (such as fans) with dust particles. The physical movement of paper through the paper path is also hard on springs, rollers, and picker mechanisms, which wear down and need periodic cleaning and replacement. The more complex the paper path, the more likely the paper will jam. Anyone who has ever tried to use the office copier knows that duplexing is a common cause of paper jams, as the paper position must be flipped to expose the second side to the print head.

Parts unrelated to printing such as PCBs, memory, and liquid crystal display (LCD) monitors in use as operator displays have problems that are typical of other products using the same technologies. Consumables (ink or toner) make their own contributions to failure with leaks and globs that jam print heads.

As a result, most printers still have standards for routine "Preventative" Maintenance (PM). No other equipment in computing has retained the need for PM to the same degree as print and paper handling equipment. Service plans frequently include some form of cleaning on a regular basis along with periodic replacements of major components such as drums, fusers, and rollers as part of a major service event. Comparisons of cost per page are deliberately difficult to calculate as various OEMs attempt to differentiate their products without losing the profit potential of consumables and service kits.

Printers are often accompanied by paper handling equipment designed to improve the efficiency of the printer and reduce the costs of paper by buying in bulk. Large systems that pre- and postprocess high-speed laser printers are also highly mechanical and subject to wear. Digital presses are another converged technology where image processing is done on the computer and the printing press tightly integrated but still highly mechanical. A high degree of interaction between the equipment owner and the maintenance vendor should be expected for all paper-processing products.

NETWORKING DEVICES

Networking equipment includes both the equipment supporting data communications as well as equipment formerly used only for voice communications. Voice and data communications equipment is largely converged technology riding the Internet for both voice and data. The remaining voice only (PBX) systems are expiring fast and are being replaced by systems based on PC system technology.

Data networking equipment is now hosting voice as well as data. The vendors of converged voice and data, predominantly CISCO, have made use of existing data networks and prospered by the proliferation of wireless systems at the front end of high-speed fiber optic networks.

The equipment layers behind any network—be it the topology (network design), a LAN, WAN, or other design—have few physical differences from other assembled electronics other than interfaces or cabling. Further, there is little physical component difference between a server and a router or switch, as the componentry is similar.

The diagram in Figure 12.2 highlights the relationship between the user-side devices (of whatever scale), the task-specific security layer (specialized "black box" devices), and the communications equipment itself. For buyers the most difficult assets to control for a lifecycle are those with a "security" function, since the products are not yet commoditized and the OEM is able to command control of the entire lifecycle. It would be extremely unusual for a corporate buyer to drop support on a security device, as the risk of job loss in the event of any "breach" would be obvious.

Because the downtime associated with a network outage is so clearly costly, there has been greater pressure placed on networking vendors than on data center processors to provide highly reliable products. Managers have demanded highly reliable devices for both interior (data center) located equipment as well as exterior (particularly weather exposed) locations. The result has been positive, as the documented failure rate of datacom/telecom equipment has been significantly better than with server cousins.[3]

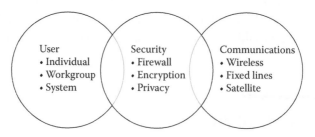

FIGURE 12.2
Network attached devices.

DISPLAYS

The original display device was not the cathode ray tube (CRT) but the teletype printer. Both have gone to the scrap heap of history and been replaced predominantly by LCD (liquid crystal display). The LCD is lighter and pulls less power than the CRT. The combinations of lighter weight and power costs have allowed LCDs to be larger and brighter in both fixed and mobile applications across a broad range of uses.

Some CRTs remain in use as the equipment and operating systems that they are attached to are not capable of adding the drivers for newer products. The largest problem with the remaining extant CRTs is screen burn, caused by the repeated display of the same information, such as a data entry form, that literally etches the inside of the glass. (The original screensavers were designed to prevent such etching.) Users needing replacements for any CRT at this time are challenged to find replacement parts. It has been at least a decade since anyone wanted to stock and sell refurbished CRTs, for the simple reason that few would survive the transportation to the buyer.

CRT technology is now treated as hazardous materials for recycling purposes. The interior of the glass is coated with heavy metals that must be separated from the rest of the unit and processed separately. Newer manufacturing technology has avoided the use of toxic products making the recycling of "e-waste" less difficult.

The dominant display product in use at this time is the LCD. It is expected that the light emitting diode (LED) monitor will replace it as manufacturing costs decline. The support problems of LCDs will last for many years beyond the adoption of a replacement product if the CRT experience is any guide.

LCDs are expensive to manufacture, although over time competition has reduced the price line to consumers. The manufacturing process of LCDs is not without defects and challenges. For a number of years the *yield* on the production of LCDs was as low as 50%.

Manufacturers have struggled to control the number of dead pixels in a display, as the underlying transistors cannot be easily repaired. The only cure for dead pixels is a unit replacement, so most manufacturers have a quality control process to spot defects before integration into products. Once assembled into a product, the active-matrix LCD (today's dominant technology) does not have a light source, so screen visibility is provided by a fluorescent light powered by an inverter that pulls current from the main power supply or battery. LCDs can fail on their own, but so too can inverters, cables, and lights. Anecdotal reports by specialist companies confirm that inverters, not LCDs, are the number one cause of LCD failure. This is fortunate, since an inverter is a relatively minor repair. (The major cause of inverter failure is the capacitor.)

Products without a need for high-resolution graphics, fast refresh rates, or even color can use passive LCDs from older and simpler designs and lower manufacturing costs. Such LCDs will probably have lower failure rates because they have fewer parts and therefore fewer points of failure. However, the nature of the supply chain may focus so heavily on meeting the demand for active LCDs that passive LCDs become scarce and expensive.

ACCESSORIES: CABINETS, FRAMES, AND CONNECTORS

Systems are linked together with cabling. These elements are not without flaws or service issues. The Institute of Electrical and Electronics Engineers (IEEE) and various safety agencies work closely on specifications so that internal or external cabling is safe, but that does not by itself prevent failure. Cables can be accidentally crimped leading to interface failure. Bulky cables can interfere with air flow, as with ribbon cables inside small cases. Poorly shielded cables can cause interference with other electronics. Some cables are rated only for short distances, so repeaters and signal boosters are needed to extend cable length. External cables often detach so easily that the first questions asked by a help desk contact is almost invariably "Did you check that the cables are securely attached?"

Other common accessories, such as external keyboards and mice, have service issues of their own. The low cost of the item often dictates that a failure is resolved with a replacement. Even if these items could be repaired, the logistical costs of handling such low value items tend to cause owners to treat them as worn out pencils and they are discarded. Many organizations do not even consider keyboards and mice as part of the configuration for purposes of depreciation or asset tracking. They have so little value that is it more costly to track them than to lose them.

Cabinets, Frames, and Connectors

There is an important distinction between devices with internal logic and electronics, and furniture. A rack is an unintelligent device that holds standard dimension equipment of any brand. A rack does not have a maintenance fee because it has no electronics.

Frames and enclosures, on the other hand, are OEM specific usually providing a common power supply, network attachments, and potentially cache memory or controller functions. These products are sold with maintenance contracts because they include parts that can fail. Some frames or enclosures are specifically designed to house components exclusively made by the OEM within the frame. Others, such as some brands of robotic tape libraries, are designed to accommodate a wider variety of attached devices.

Most of the service issues associated with frames and enclosures are difficult to associate with the correct part because of the confusing way that they can be called out in asset management and service desk systems. If a trouble report is put out for a blade server, but the problem is a bad power supply in the blade enclosure, the service desk tracking system may not pick up the differentiation. The service vendor may then be unprepared to bring the right parts, leading to a Service Level Agreement compliance issue, leading to an argument.

By comparison, individual servers mounted in racks cause little trouble. Servers are individually managed, individually serialized, and can be transported from rack to rack without any complications.

SUMMARY

New settings for technology products will likely need repair for the same reasons as those in other settings. The product may be more or less reliable due to environmental conditions, but the basic design and the quality of manufacturing does not change. Before buying a "new" product, buyers can investigate the repair needs of the same components in other settings before agreeing on a purchase.

NOTES

1. Martin K. Anslem, "Failure Analysis: Lessons Learned," http://www3.uic.com/wcms/ images2.nsf/(GraphicLib)/FA_Class_2012_pres.pdf/$File/FA_Class_2012_pres.pdf. This study from Universal Instruments (a provider of testing tools) includes hundreds of photographs of failures of PCBs with accompanying text.
2. For a brief but technical explanation, see "Topic 5: Integrated Circuits," http://wps. aw.com/wps/media/objects/877/898586/topics/topic05.pdf.
3. TekTrakker analysis from 2009 showed that several common models of CISCO networking equipment posted failure rates of about 20 years (240 months) mean time between failure. At the same time, common server models in use were posting failure rates of roughly 80 months mean time between failure, showing that the focus on high reliability for networking uses was producing models with superior failure profiles.

Section IV

Controlling Product Life

13

Building Blocks of Using the Machine: Software Layers

INTRODUCTION

This chapter breaks down the three major types of software, all of which have their own licensing traditions and terms and condition issues. Each of these layers interacts with the layer below and ultimately the hardware to provide a complete system. The chapter deals with each of the layers in the series from the top to the bottom. The organization of the chapter is deliberate to reinforce that maintenance and support of the upper layer does not maintain or support the layers below.

SOFTWARE LAYERS

Figure 13.1 shows that corrections to the application layer do not flow downward into the operating system. Nor does maintenance, and support of the operating system (OS) does not correct flaws in the machine code (bottom) layer. In turn, the machine code layer does not fix physical problems but does repair logic errors that may impede function.

Keeping these limitations in mind is key to understanding which parties must interact, or not, to provide repair and maintenance.

TOP LAYER CONTENT AND MEDIA

"Content" refers to the books, music, games, movies, and so forth, that are provided in digital format but are not so much software as they are discrete

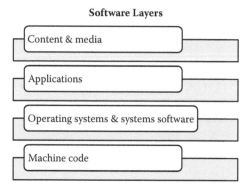

FIGURE 13.1
Diagram of software layers.

products. In an e-reader, such as Kindle, the content is the book, and the software that allows the book to be read is part of the machine code of the Kindle. Many similar products exist that appear to be hardware, such as game stations, but are not intended to be repurposed even though the hardware itself is useful for other tasks. This design approach in turn greatly limits the use of the hardware, including repair, and forces the owner to deal exclusively with the original equipment manufacturer (OEM) for support and maintenance. It is not so much that the game station or e-reader OEM is against consumer repair, but by granting access to the machine at the level needed to ease repair, the OEM would also be making it far easier for consumers to play pirated materials.

Hardware designers for content players make extensive use of access blocks to the machine code layer so as to prevent piracy of content. Although the makers of products intended to use content are aware that their product designs are limiting owners in terms of repair, their primary business interest is elsewhere. It remains up to the user before purchasing the equipment to determine if these limitations are acceptable.

Unfortunately, most users do not have the opportunity to review the terms and conditions prior to purchase. Many content-using products are sold with "in-the-box" licenses that are not available for review prior to purchase. Another common practice is to provide a license acceptance checkbox during installation or installation cannot proceed. While this may be considered adequate notice for legal purposes, it is not practical for purchases outside of the consumer market.

Software support, if there is any, is provided only by the OEM for the content player or only by the content provider. Content itself is completely protected by copyright law and there is no market for independent *support* of someone else's original material. If the game does not play correctly, only the game designer can fix the source code.

APPLICATIONS

In-the-Box Applications

In-the-box applications are the applications sold by retailers directly to consumers in boxlike packages. The contents of the box include some media, an instruction guide, and a lot of empty space inside the cardboard cover. There is nothing that a buyer can do to modify or adapt the application provided, except with whatever limited customizations are included with the boxed product.

It is common that in-the-box applications are no longer delivered in a physical form at all and are directly downloaded from the vendor. The terms and conditions of the purchase remain the same with or without the packaging. Many in-the-box products do not provide the details of the terms and conditions in advance of product purchase, which can create a problem of contractual fairness. Good contracts should allow for disclosure of terms and conditions prior to sale, rather than after the seal on the packaging is broken.

Support for in-the-box applications is limited and may consist only of a toll-free number or a Web site URL for tech support. It is expected in this end of the industry that problem resolution is contracted separately on a per instance basis. The original retailer has no involvement in the support process at all and usually rejects any attempts to return opened boxes. Whatever recourse the buyer might have is between the buyer and the application owner.

Commercially Developed Applications

The most well understood category of business software applications are purchased to perform specific functions. Regardless of the simplicity or complexity of the application, the developer of the software is the owner and unless they offer their software for free, as in "open domain" products, then the only source of support is the application developer.

Applications are designed specifically to run on a particular OS and use only the instruction set provided with the OS. As a result, applications are not immediately transportable across different operating systems and separate versions must be developed to run on different platforms. Similarly, new versions of the operating systems frequently require extensive updating on the part of the application provider in order to remain compatible.

Version levels of OS are extremely important to the application developer. Applications have version levels that match to different iterations of the OS as well as different iterations of patches and fixes for the application itself. Version level control for applications is an important element of asset management and problem determination, as version level conflicts are a common cause of software problems.

The hardware platform is of far less concern to the application developer than the operating system. If the hardware is capable of running the operating system, the application should work even if it works slowly or poorly. Some applications run far more efficiently on particular operating systems, which in turn may be optimized for particular hardware.

One of the observable consequences of new application development is that assumptions are made that the underlying hardware is at the robust end of the available technology choices for equipment. If the platform is not using the latest and greatest of available components, the application may be such a large resource hog that the machine can barely execute the code.

The desire to run new but resource-intensive applications or newer versions of applications is a primary driver of new hardware purchases.

There is little incentive to write *tight* code for applications, as hardware capabilities remain on the capacity curve predicted by Moore's law. Much that is written for applications could be condensed and made more efficient, but the availability of cheaper and faster components does not justify the investment in more painstaking development.

Patch management is an essential part of supporting the suite of applications installed by the end user. It is likely that some applications will not function correctly in the presence of other applications and the only way to begin to ascertain these conflicts is through careful record keeping. It would be unusual for any application provider to be able to fully test all possible combinations of products in advance of product distribution. The permutations of possible conflicts are vast and may be functionally infinite.

The best that most application vendors can do when it comes to anticipating product conflicts is to test the most common combinations and hope for the best. The more exotic the mix of products in use, the less likely that the application developer will have tested these combinations. The wider the variety of combinations, the more likely that the problem cannot be easily replicated and resolved.

Homegrown Applications

Most applications in the early days of computing were entirely homegrown. Directly employed staffs of programmers built custom applications from the ground up, often using languages no longer in vogue, such as Assembler and COBOL. Many of those original applications are still running in large enterprises under layers of workarounds because the source code is missing or undocumented. It is also a known problem that there are few programmers alive skilled in the older and unsupported languages.

The lack of source code versions of very old products has created a niche market in emulators that allow very old code to be run in more modern settings. Running under emulation is not bad per se, but the lack of source code is more problematic since very old software simply cannot be updated to perform new functions.

The painful experience of trying to keep homegrown applications updated and working have led most enterprises to avoid building custom applications and turning instead to ready-made products. Buying a fully built and professionally supported application is far more efficient than starting from scratch although a custom product would be more desirable.

Customized Applications

Customization of applications is a common requirement. There are few applications that can serve the variable and complex needs of an organization *out of the box*. Expenses to customize applications often grow exponentially with both changes in specifications and project creep as completely new requirements are added. The best way to keep costs under control is to do as little customization as possible.

Support in the customized environment is also likely more expensive and limited than in the stock product. Only a handful of programmers employed by the software vendor will have the insight into the routines added to the customization and they become ever more valuable as time goes by. If these people switch employers, the user may be stranded.

Authorship of customization is also an issue. Some customizations are so vast that they become a separate product onto themselves, and the rights to market the resulting code must be negotiated. The Statement of Work covering the customizations must include clear parameters, mutually agreed upon, that cover how support of any customizations will be provided and how they will be charged for in the future.

Horizontal and Vertical Applications

Within the application development community there are distinctions made between widely useful applications (horizontal) and applications with a narrow focus (vertical). This distinction is the basis of many business partner authorizations where the application license sale is linked to the provisioning of products and services made by others.

For example, the IBM AS/400 (now the I series) was sold by thousands of authorized business partners as the platform that was needed to run the specific software application. IBM limited these authorizations to software systems that were narrow (vertical) so that there would be little or no competition for the product sale since the combination of products was essentially turnkey.

The sale of verticals typically included the sale of the hardware platform and all of the operating system licenses that supported the application. Along with the hardware and OS licenses, the partner would normally add additional hardware break–fix (maintenance) contracts, installation and customization work, and then offer a lease package for the whole project. (The known structure of these arrangements requires that the packaged price is a sum of many parts and can be disclosed separately at the request of the client.)

Application vendors with horizontal (widely useful) applications such as accounting software or payroll are less frequently authorized to partners because the generalized product sale can be more competitive with other solutions from other vendors and may expose the transaction to a price battle. The application developer is more likely to use their own sales force to sell the software separately from the hardware so that the application decision is made independent of the platform.

Defect support for these vertical applications lies entirely with the author/licensor of the application but often includes the business partner in the support process. It is typical that the partner is required to provide the first line of support (the initial trouble call) as part of their partnership agreement. Only after the first level of support has triaged the call is the call for hardware service (maintenance) or software support (the operating system) made.

Mobile "Apps"

Although widely known as "Apps," mobile apps software products are not "applications" in the same sense of business-class applications, particularly in terms of support. The buyer of a mobile app does not get any warranty or support with the nominal purchase price. The app either works satisfactorily or the buyer does not use it. With hundreds of thousands of apps available for a wide variety of mobile platforms, there is tremendous competition and innovation coming forward from this space, almost none of which is formally supported.

PROGRAMMING LANGUAGES

The choice of programming language is up to the developer, but the buyer of developed products should consider the downstream ramifications of that choice. Applications need support, and programmers skilled in the use of the programming language are required. If the language is common, the talent pool for experienced programmers is larger than for a very esoteric language. Developers can go out of business leaving successors with unintelligible source code.

Programming languages can be proprietary or "open source." Even within the open-source community there are different types of licenses that limit how products built using the language can be promoted. There are proprietary languages that allow customization by individual users but do not allow proliferation or resale of customizations.

Languages described as "Higher Level" are intended to make repetitive tasks simpler and programming instructions more easily understood and documented by the user. In order to execute the program, the higher-level language must be processed to include all the hidden subroutines into machine-readable code. This process is called "compiling" or "interpreting." Most common business computing is programmed using higher-level languages as shown in Figure 13.2 provided by Tiobe.com, which tracks the use of languages.[1]

TIOBE Programming Language Rankings

2002	2013
☐ Java	☐ C
☐ C	☐ Java
☐ C++	☐ C++
☐ (Visual) Basic	☐ Objective-C
☐ PHP	☐ PHP
☐ JavaScript	☐ C#
☐ Python	☐ (Visual) Basic
☐ C#	☐ Python
☐ Transact-SQL	☐ JavaScript
☐ Objective-C	☐ Transact-SQL

FIGURE 13.2

TIOBE programming language rankings. (From www.tiobe.com.)

Programming languages are categorized by how they use instructions and the general purpose of the language. As shown in Figure 13.2, Object-Oriented Programming (OOP) is popular because it allows the manipulation of packets of data and functions as "Objects," so the package can be used in different ways without reprogramming. The ability to constantly reuse objects rather than continually redescribe them reduces the potential for errors in programming, which is the major advantage other than speed.

Procedural programming is similar but less flexible than OOP in that prescripted procedures are available to the programmer to call on for common routines.[2] The procedures simplify repetitive steps, which can be functions other than mathematical. The programmer still has to express each step in the program specifically, without the advantage of being able to move "objects" in chunks. Many older procedural languages have been updated to allow OOP programming, but as with any emulation, the results are not as clean as for a language specifically designed for the purpose.[3] Functional and logical programming languages are still in use for mathematical purposes, which was always their original intent.

Regardless of category or name of the language, the main concern for buyers of software packages is support. If the developer goes out of business, many contracts specify that the buyer is entitled to the source code, which is important, but if the source code is poorly documented, then having the source code is of little value. Each software purchase is a trade-off between buying easily supported code written in higher-level languages or more efficient code written in less readily accessible languages.

OPERATING SYSTEMS

There are many more vendors of operating systems than is commonly appreciated. Some products, particularly those with low programmability or a single function (such as a cable TV box or the Engine Control Module [ECM] in a vehicle), are delivered with most of the operating system on the chip. Some of these OS elements are proprietary, but most are considered part of the hardware. When corrections are needed for these products, the patches provided are rarely differentiated between hardware (microcode and firmware) or software (OS) patches. The details of such corrections are irrelevant to the end user since they do not separately contract for OS and hardware maintenance separately.

Commercially available operating systems, such as Microsoft Windows or IBM z/OS, are proprietary works protected by copyright. Users of these systems acquire and license the OS separately from the hardware, and the only source of support (patches and fixes) is the owner of the software.

Open domain operating systems, such as Unix and Linux, are ostensibly free, but the basic code is so primitive that most users purchase an independently supported version, such as HP-UX or Red Hat for Linux. The vendors of the modified systems often copyright their intellectual property to the degree that it can be done. Once committed to the supported platform, the user is in much the same position for updates and patches as from the proprietary software vendor.

Patching

For products where the support of the operating system is separate from the support of the hardware, differentiation is important in order to associate the problems experienced with the correct contract. Some vendors bundle all patches together in a single release of updates. This is sold as a convenience to the user but hides the true nature of the patches. This is not an accident. The bundling of internal (noncommercial) OS patches is also one of the key methods that vendors employ to strong-arm end users into buying both hardware maintenance and software maintenance from the same vendor. If there were not such an artificial tie-in, the patching of hardware issues would belong with the hardware sale as defect support, and the continued updating of the software properly belongs with the software license.

Peripheral Drivers

Drivers for peripherals and products now integrated in the frame but technically separate (such as controller cards, graphics processors, disk and tape drives) are usually provided by the OS vendor as a convenience for the user. The drivers themselves are created by the vendor of the peripheral and are always available separately as downloads (or on media provided by the peripheral vendor). It is not usually necessary to buy the new version of the OS in order to attach new peripherals, unless the machine vendor specifically thwarts the process by denying support for something other than the peripheral. This is a negotiable point, particularly if the OS vendor has not created the driver but merely passed it along.

Security Patches

Patches with high importance are often labeled "Security Patches" to get the attention of the owner to apply them quickly rather than wait. These patches are not upgrades or enhancements; they are correcting known defects in the product that would otherwise impact correct operations. By using the high urgency label of "Security Patch," the vendor also subtly (if a hammer is considered subtle) pressures the end user to continue to contract with the vendor for *support* years in the future in order to have access to such important patches.

Some healthy skepticism must be adopted to avoid being caught up in the security patch noose. Most patches are created in response to problems that arise early in the product distribution cycle. Once products are stabilized, they are rarely modified as both the hardware vendor and the software vendor will have moved to focus their attention on the next actual revenue-driving enhancement. Vendors do not fix products that are not broken nor do they provide enhancements for free.

When faced with a contracting decision regarding gaining access to *security patches* through the OEM, users do well to ask the vendor to summarize the purpose of all patches over the last year of activity for that particular product. This request should reveal the true nature of the patches and the user can then judge if the price for access is appropriate.

No matter what the purpose, patching is always a bad thing. It is easy to make errors while patching, as the guts of the operating system are exposed and altered. Patches themselves may be introducing yet other errors, and some patches are applied that create additional conflicts with other machines, configurations, or OS code.

Patching is also an indication that the product was not fully tested prior to shipment. The more patching that is needed, the less confidence the user should have in the delivery promises made by the vendor. I have toyed with the idea of collecting patch frequency data to be able to monitor products as to their stability as a guide to quality. The idea for development was dropped since, armed with this information, most users would still unfortunately be stuck with their existing vendors. Instead, I suggest that end users monitor their patching frequency and tie it to a contract so that excessive patching can result in some form of rebate to offset the increased costs and risks.

SYSTEMS SOFTWARE (ACCESSORIES AND UTILITIES)

There are products designed to perform specific functions that are not included within a licensed OS. Many times, the success of these ad hoc products becomes the guide for enhancements to the licensed OS, much to the dismay of the original developer. Many of the functions now common within the OS used to be separate products that were developed to fill needs that were not provided by the hardware vendor. The history of how these functions were developed provides some insight into how deeply integrated and powerfully dangerous these products can be.

In the early days of IBM mainframe computing, the operating systems were free but also quite limited. For example, for a long time one could run only one job at a time, such as print a report. There was no such thing as printing in "background." It was obvious that more functionality was needed and a marketplace quickly developed for imaginative programs that enhanced operation. Some of the genius geeks of the era became wealthy distributing their innovations. Some of the companies formed to serve these functions are still in business today, including Computer Associates (CA).

Executing these functions required that the OS itself be modified. As IBM was providing its OS in the open domain (a concept that had not yet been invented), it was perfectly legal and reasonably common. Systems programmers readily shared invented routines (a precursor of "Shareware") and were comfortable tweaking the OS itself both to apply needed patches and to augment the functionality. The systems programmer was a god among geeks and had the skills to command exceptional pay. It was no surprise that organizations somewhat welcomed a chargeable OS with supported patches rather than to continue to allow the systems programmer to have his (almost all men) fingers in almost everything.

For example, some products trapped every request for an input/output operation in order to redirect every request for data to the correct location. When these products crashed, they tended to take down the entire operating system, so prompt problem determination and resolution were essential. There was no tolerance for failure and heads rolled, both at the end-user location and with the vendor, if preventable errors were permitted.

Such modifications today would be impossible. The vendors of the proprietary OS would block access and almost certainly refuse to support their OS if there were any outside code installed. Managers, already risk adverse, would never put their jobs on the line to experiment with novel approaches. The result has been an almost institutional acceptance that the OS is sacrosanct and cannot be touched without inviting the wrath of the Almighty. In turn, end users have lost nearly all competition for quality and price when it comes to the operating system. Once installed, it is nearly impossible to replace or to command quality from the OS vendor by switching vendors.

Outside the mainframe data center, the independent systems software industry took hold in a very large way with first supporting Unix and then Linux. All the companies that provide supported versions of open domain products are in the systems software industry.

All applications are written with the underlying OS in mind even if the OS is not separately licensed (as with the Apple IOS). Although the end user may not see the differences between a Mac version and a Windows version of an application, the way the systems work on the inside are very different. Different versions of hardware also impact how applications operate, such as being able to use the power of a 64-bit machine for functions impossible on a narrower interface. Therefore, purchasing/licensing one version of a product for use on a particular platform is not likely to operate on a different platform unless the author specifically recommends it.

Every electronic product that runs an application has an OS. It is the OS that handles all the internal details of moving data from storage to memory to processors and back. Operating systems execute all the instructions of the application system. The OS provides a standard set of instructions for the application programmer to call upon. Without these standards, applications would be far less easily built and be far more difficult to support.

All operating systems are systems software, but there are accessory and specialty products also known as systems software. All systems software is both hardware and OS dependent.

License Forms

The developer licenses systems software. Some OS products are technically free (also known as open domain), but the free version has no support. It is possible to run on free software, but this is extremely difficult to do without the involvement of a tremendously talented and experienced staff that can write in assembly language and diagnose the myriad problems of connecting new devices to old systems. For example, UNIX as a free OS dates back to the 1960s[4] but is functionally unsuitable for today's equipment suite.

Supported versions of open domain operating systems include famous products such as HP-UX, Solaris, and AIX-supported versions of the original free UNIX software; and Red Hat, a supported version of the original free Linux software. Once licensed, the providers of the licensed version of the product are the only source of support.

The glue holding the customer to a specific type of operating system is driven by the choice of an application and accessory products. If the organization is running its business on an application that only runs on a particular operating system, the user will have no choice but to invest in the systems (both OS and hardware) that enable the application. In some cases, as with "appliance"-type products, the application is inseparable from both the OS and the hardware. Users of this type of system have no competitive options to control vendor terms and conditions other than competitive selection and careful postwarranty planning.

Systems Software Support

It is common for the initial release of systems software to need repairs as field use reveals incompatibilities and conflicts that can be resolved in a patch. Such fixes are variously described as patches, fixes, updates, security updates, service packs, and similar words for repair or adjustment. One has only to have owned a Windows machine to have noticed the constant alerts to security updates and *service packs* to be familiar with the concept.

Just as fixes to application software do not make corrections to the underlying operating systems software, patches to the licensed systems software product do not extend down to the machine code level. Nor is it possible for patches to the systems software product to repair physically broken hardware.

Patches to systems software are made by the developer to fix errors. The sequence in which patches are created is material, because some patches must be applied in a particular order. It is also possible that some patches undo prior patches and others cannot be undone. Because there can be multiple patches available to resolve multiple known problems, users need to have ready access to OS patches as well as version-level tracking at the serial number level. Patch management is a common driver for implementation of sophisticated asset management systems.

MACHINE PROGRAMMING

Managing today's systems does not require the ability to program at the machine level, but understanding the way that electronic equipment is built allows insight into vendor behavior that is advantageous. It is easy to be intimidated by threats of dropping support, as an example, when the educated manager will understand intuitively how valid or preposterous such threats might be.

At the most basic level, all electronics are combinations of on–off switches known as bits. Each bit is either on or off in exactly the same manner as a household light switch. Arrays of 8 bits are known as bytes. The combination of bits within a byte expresses numbers in what is known as binary (mathematically base 2). It was easy to see bits and bytes in action with the computers built from vacuum tubes as the tubes lit up as various computations were made. However, it is difficult to visualize large numbers in binary representation, so a more human-friendly set of representations was created and became what we know as "Hex."

Hexadecimal (hex) is mathematically base 16 and uses digits 0 to 9 plus additional letters, usually A to F, to represent information that can be manipulated by the machine in binary. There are some variants but most commercial computing remains based in hex. The use of hex also explains the sizes of memory and storage systems as manufacturers created machines to deliver even numbered combinations of addresses, thus we have 32 bit, 64 bit, and so on, all of which are combinations of 16 at a time and not decimally organized systems of, say, 50 or 100 bit machines.

All data is converted into hex in order to be manipulated within the machine. Addresses and locations in storage are arranged in hex. Manipulations of data, such as adding two values together, are called for by the programmer using the instruction set created by the hardware manufacturer known as Assembler.

Machine Languages, Programmable Language Code, Assembler, and More

Programming of machine processes is based on an instruction set provided by the hardware designer. Each of the instructions is provided to perform simple but essential instructions such as moving data from memory into registers or moving the results of computations into storage.

There are multiple names for such code. Some manufacturers use the phrase "Assembler," others use "Programmable Logic Code (PLC)," and still others use the term "Machine Code," but the function is the same.

The instruction set for available programming functions often includes common routines in order to expedite processing. Different types of machines are provided with instruction sets that may be more suited to math than other machines. It is for this reason that certain brands of computers are used more frequently in specialty settings than others.

Machine code routines perform such things as order the operations of starting up, and reading and writing from storage. As equipment has become increasingly sophisticated, it has been possible to include more instructions on the chip itself and also to provide an update function to make corrections as flaws are identified.

The process of error correction to machine code or PLC is commonly known as "Patching" and specifically patching of microcode or firmware. (This is an area where hardware and software are inextricably entwined.) There are considerable contractual distortions forced on end users for the defect support of machine code, notably the tying of support contracts for operating systems support to the entirely separate obligation of hardware (machine code) defect support. Machine code instructions are entirely separate but poorly understood as such.

It was not always the case that machine code was manufactured into the chip itself. For a long time the instruction set provided by the manufacturer was known as Assembler, a fundamental programming language that allowed the programmer to manipulate the sequence of events at the machine level. As might be guessed, assembly programming was difficult, exacting, and time consuming.

Assembly programming was essential due to the limited speed, storage, and memory capabilities of the machines produced in the 1960s through the 1980s. If code was not tight, meaning that the fewest possible instructions were used to execute a routine, then the programs would execute poorly. Errors in Assembler routines were quite catastrophic and generally caused the machine (and the job that was being executed) to abort (ABEND) and the machine would need to be restarted. Tight code was highly valued. As a result, most systems software, including operating systems and utilities, were written in Assembler (or the manufacturer equivalent).

Higher-level languages, such as BASIC, FORTRAN, or COBOL, were created to ease the burden of programming in Assembler. It was no longer necessary for a programmer to know how memory was utilized within the machine or how to move data within the machine, but instead write programs that had a business or scientific purpose. The process of "Compiling" programs written in a higher-level program functioned to turn the final product into the assembler language the machine would execute. Use of higher-level languages mirrored the exponential improvements in hardware speed and storage capabilities.

The vast capabilities of today's chips and storage have all but eliminated the need to write tight code. This is advantageous for hardware manufacturers seeking to sell increasingly powerful machines. So long as the next iteration of a chip is required to support the requirements of a robust, but perhaps excessively lumbering operating system, users will continue to be compelled to make costly hardware changes.

ACCOUNTING FOR SOFTWARE

Buyers of technology equipment, regardless of purpose, expect to capitalize the purchase. The hardware portion of the equipment is depreciated, and the software portion is treated as a license and is not depreciated. Lessors and lenders that may be involved hold a security interest in the equipment but not the software. Owners of hardware can borrow against the value of the asset and the total value of tangible assets is part of the financial strength of the organization. The breakdown of tangible assets (hardware) from intangible licenses (software) is fundamental to financial accounting.

Understanding the relationship between hardware and software is the lynchpin between OEM domination of hardware service contracts and end-user flexibility. It should not matter if the device in question is an automobile (with significant electronics), a mobile phone, or a server. The differentiation between hardware and software is fundamental to the long-term value of the item.

Ownership of Embedded Software

The dividing line between tangible and transferrable assets and intangible licenses is the treatment of machine code. Autos have high residual values and long useful lives because the digital parts are sold inclusive of embedded software. Owners can repair equipment and keep it in service without concern about violating license agreements. Owners can sell used vehicles as whole machines, and buyers do not have to bargain with the manufacturer for the right to use whatever embedded software was provided as part of the original purchase. Vehicles become scrap only after they have been physically destroyed (as in a major accident) or become too costly to repair. The automotive aftermarket is vibrant both in terms of stripping damaged cars of valuable parts, and in recycling metals and plastics for profit.

None of this is the case if embedded software is licensed, which has become common in computing. Information technology (IT) equipment is sold inclusive of licenses that cannot be transferred from the lowest level to the highest. Nontransferrable machine code licenses are being promoted by the OEM even when the machine will not operate whatsoever without the code. The license for machine code is then used to prevent the transfer of tangible assets into the secondary market.

This has the same impact on used computer sales as it would on used vehicles. The used car buyer would not buy a car without the machine code, and must therefore negotiate not only with the seller of the car but also with the original manufacturer to acquire the rights to use the car. If the license transfer were nominal and timely, such as an odometer statement, the used market would still operate. However, if the OEM decided to charge hefty prices, or demand upgrades, or demand associated service contracts, or even failed to approve the transfer in a timely fashion, each such requirement would rapidly degrade or destroy the used market.

IT products are not inherently valuable as raw materials. The frame is lightweight and the components worthless as scrap metal or plastic. If OEMs do not allow legal repair of their designs in the secondary market, there is no secondary market for parts. Although vehicles enter the scrap market over decades, OEMs have the ability to cause their products to be scrap the moment the ink is dry on the purchase agreement.

From a lending perspective, any terms that reduce the value of the asset or create difficulties with resale, immediately diminish the appetite for lenders or lessors to be involved with the equipment. One solid reason to avoid such equipment is a default. Lenders do not want to lend against equipment they cannot repossess and sell to recover their remaining investment. Lenders stop lending, or lend at rates more in line with unsecured credit, if the equipment cannot be easily restored to service or resold.

Licensed Software and Residual Value

Licensed software support terms and conditions can also kill residual value. If the license renewal or license maintenance costs are made high for machines coming to the end of term, the end user will face a fabricated situation intended to make a new purchase more attractive than keeping the older equipment. The OEM wins a new sale, but the user never gets control of the asset and cannot take advantage of the substantial (often over 90%) decline in value in a secondary transaction.

A case in point is the disk array. Providers of these products understand that their end users wish to lease the equipment. The end user is enticed with a single lease payment inclusive of all hardware and maintenance along with all software and associated maintenance for the selected lease term. At the end of the term, the renewal or maintenance costs for the license portion alone is made deliberately unattractive so that the proposal for the total monthly cost of the replacement product is always more attractive. In this way, using software license pricing, the OEM has created the situation where there is no residual value for these products. In turn, because the OEM has not only crushed its own residual value, lenders will not take the equipment as security in the transaction. The OEM wins not only the constant churning of more equipment sales but also controls all financing, trade in, and the secondary market.

The hardware array itself can be broken into parts for secondary market use, but no whole machines are traded as a consequence of OEM license transfer or renewal fees. OEM lease terms and conditions prohibit filling up empty arrays with used disk drives, so that the only market for used equipment is the very few parts that can be absorbed by independent repair companies. No whole machines are traded in the United States because the licenses are too costly compared with other options.

End users who are frustrated with being manipulated in this manner can only vote with their purchase orders and cease dealing with companies that engage in these practices. There are always options including operating on frozen releases and independent service.

Purchase, Rental, and Leasing of Software

Software licenses for OS, systems software, and applications software are never purchased. The intellectual property (IP) remains that of the software provider and copyright law protects the IP. Perpetual agreements are intended to provide some of the price stability of a purchase, but the license itself never transfers. Perpetual or purchased software is also subject to maintenance fees for support, which are traditionally linked to the original purchase price as a percentage.

Rental of software is a fully viable technique whereby the vendor is paid periodically to provide both the license and support. The rental agreement is common to Software as a Service (SaaS) products and Cloud products where the software is never installed at the end-user facility.

Longer-term agreements for licenses are typically called leases, but the agreement is much more like a rental than a hardware lease. The issue is one of transferability. A software license cannot be transferred to another owner so it has no residual value and cannot be used as part of an asset base for credit support.

Lenders treat stand-alone software leases as a pure credit decision. Without any collateral backstopping the value of the license in the event of default (such as a bankruptcy), lenders always charge an interest rate premium for intangible financing. Under these circumstances, software vendors have a significant financial advantage over banks in offering multiyear "lease" agreements because they have very low capital risk and can unilaterally make decisions for the transfer of licenses in the event of a bankruptcy.

Many OEMs selling hardware will set up lease-like financing for their OS products as a service to the buyer, but the software items are always itemized separately and calculated against higher lease rate factors. Lease agreements that do not separate software from hardware problems should be investigated for breakdowns of pricing so that end-of-term renewals do not accidentally include extensions of financing that has concluded.

End users have become accustomed to licensing applications and are generally respectful of the concept of intellectual property for applications. End users purchase applications knowing that the support is either up to them, available independently, or uniquely available from the OEM. The license agreements clearly spell out the relationship between the parties.

SUMMARY

Buyers of software licenses are mostly aware that they cannot support or modify the software without the cooperation of the licensor. Vendors of software, unless under the credible threat of a competitive displacement, are in a monopolistic position when it comes to negotiating any postpurchase services, particularly for long-term maintenance or customization.

NOTES

1. TIOBE Software (www.tiobe.com) provides additional tracking of programming languages and ranking methodology.
2. See "List of Programming Languages," Buzzle, http://www.buzzle.com/articles/list-of-programming-languages.html, for an excellent list and description of programming languages, their purposes, and categories.
3. See "Object-Oriented Programing," *Wikipedia*, http://en.wikipedia.org/wiki/Object-oriented_language, for a discussion of the history and purposes of object-oriented programming.
4. See "Unix," *Wikipedia*, http://en.wikipedia.org/wiki/Unix, for more about Unix as an open-source operating system.

14

Software Support Issues

INTRODUCTION

This chapter covers issues of contracting for software support from the inception of the purchase through product retirement or replacement. The nature of the software license agreement tightly limits options for support, so prepurchase negotiation is particularly essential.

DEVELOPER CONTROL

Regardless of the type of software involved, the developer ultimately controls how support is delivered. Licenses entirely govern the relationship between the application vendor, or the operating system (OS) vendor, and the user. This includes how the product can be used, supported, or customized. The user does not own the code, and therefore must treat licenses separately from fixed assets for accounting purposes. Even if the license is purchased as a "Perpetual" license, the ownership relationship is not the same as with tangible assets where assets can be traded in the used market. Perpetual licenses entitle only one party to the unlimited use of the code and are not commercially transferrable, at least not yet, in the United States.

There are a few common license contracting models outlined next with their relationship to support and maintenance.

Source Code

Ownership of source code puts the full burden of support and maintenance on the owner. The developer may offer to make enhancements or upgrades or fixes for a separate fee, but the owner can decide to do whatever they want without the permission or involvement of the developer.

Perpetual License

Perpetual licenses allow the user the indefinite use of the application software at a particular version level without further payment. This is what happens when we buy software in the box at a retailer: we buy a license with a limited period of "warranty support" and we can continue to play that game or use that version of a genealogy program indefinitely. The vendor of the application may offer upgrades or new versions or even customization, which is completely separate from the purchase. This is also what happens when an enterprise decides to "freeze" a release level and stop paying for support on a perpetual license.

License to Use/Right to Use

Not all applications have a purchase or perpetual option. The license to use or right to use (RTU) agreements are for a definite period of time and must be renewed or repurchased in order for the use of the product to continue. Most often the license includes all the updates, patches, fixes, and enhancements for that release for the period of the license term. Pricing for these is highly variable including prices based on the number of users, number of concurrent users, the number of transactions, the number of processors, or by serial number. Others offer licenses for the entire enterprise and some for a single user.

The details of the support services included with the license are specified and controlled by the owner of the product. Some products include tech support, others charge separately. Some offer customization and others do not. Upgrades and rights to new releases and pricing are also points of negotiation.

Rental

There are products that can be rented for short periods of time. The rental agreement usually includes all maintenance in the sense of patch availability but would not likely include extensive training, customization, or tech support. Most SaaS (Software as a Service) products are rental agreements that recur indefinitely unless stopped by a deliberate action. Pricing in a rental agreement can be changed quickly, often with just 30 days notice, unless the contract specifies otherwise. The same flexibility of cancellation on short notice applies also to the provider of the software, including the right to modify the product at any time or stop providing service.

Pay for Use

Although not a common licensing model for programming applications, the pay-for-use model is common in media and content, such as the rental of movies and group collaborative services such as videoconferencing. SaaS services can be priced and delivered as pay for use. Support of pay-for-use programming is the same as for SaaS options.

Open Domain

Products offered without royalties or license agreements are called "Open Domain." They are free to download and use indefinitely but do not include support. Popular open domain products often become the focus of support specialists who offer their version of the open product with their particular support capabilities. These support offerings are just as variable as traditional application licenses.

VERSION-LEVEL TRACKING

Tracking down the precise interaction of which versions are causing which conflicts with different operating systems is difficult work. Many times the problem is difficult to reproduce, and if the problem does not recur, the problem cannot be diagnosed. As we all know from personal experience using personal computers, the vast majority of mysterious problems disappear with a reboot. The same is also the case with more complex systems. The lack of being able to reproduce a problem is the main reason many "glitches" remain unresolved. Recurring problems are more easily resolvable than mysteries.

To try and sort through the complexity of these interactions, license management software is often used to specifically track version levels of software to the individual asset. If such data is merged with the hardware maintenance reporting for the asset, the whole support picture can be culled for trends and analysis. Unfortunately, these functions are often purchased and used separately. Unless each asset has a known configuration of both hardware and software (known as the "Software Stack"), correlations between hardware and software problems are impossible to prove.

PRODUCT AND VERSION-LEVEL CONFLICTS

Ascertaining which product (or version) does not work well with particular versions of other products is at the heart of conflicts between vendors over support. No one wants to invest resources in problem determination only to discover that some other product is at fault. And the flip side is also true: no one wants to find himself or herself at fault when a failure can be blamed on a different vendor. This is the time-honored tradition of finger-pointing (Figure 14.1).

There are no Marquis of Queensbury rules on who gets stuck doing the work. Nor are all conflicts resolvable. Some problems simply cannot be fixed. Sometimes users need to revert to older versions of products (both hardware and software) to work around a version-level or product conflict.

Manufacturers can create contracts that make the issue of version-level conflict out of the eye of the end user, but this does not mean such conflicts do not exist. Large software companies with large suites of software products can provide a single point of contact, but the work needed on the part of the end user to help with problem diagnosis is the same regardless of the number of toll-free numbers called. The only thing missing from the single-vendor point of contact and the multivendor support paradigm is putting up with hearing each vendor blame the other.

Ultimately, the vendor of the product with the most to lose in the account usually takes the lead in problem resolution. It is also most often the case that small vendors will patch their product to work with the product of the larger vendor.

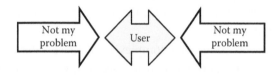

FIGURE 14.1
Finger-pointing.

Software Maintenance

The list of software maintenance contracting issues in Figure 14.2 was put together by a customer council under the Gartner Group in 2010 with an intention of spurring a code of conduct to be adopted by software vendors. The code of conduct does not appear to have been adopted, and as a result the issues remain.[1]

Software vendors are not consistent with providing a road map for their future development. Road maps are somewhat traditional for hardware vendors, most likely because the commitment to manufacturing a new version is a lengthy process involving massive capital expenditure. Not only is a new product line impossible to hide, but vendors want buyers to be willing to invest, emotionally or literally, in their platform planning. Perhaps because of the risk of having ideas copied ahead of a new product announcement, the software developer community, outside of the operating system developers, is tight-lipped and low on public planning.

Neither hardware nor software maintenance is offered based on need, which makes for considerable overspending for both types of support. The key to relating support to need is for buyers, owners, and licensors to start measuring need, and linking contracts to service delivery based on need. Without measurements and accountability, need-based service and support will not happen. For more on the subject, see Chapter 16.

Users are vulnerable to maintenance fees that are variously unfair, predatory, too high, or unreasonable. This is a problem of original equipment manufacturer (OEM) control, particularly with very large vendors with little or no credible competition. If the user cannot realistically move to a different product, the OEM (developer) has a de-facto monopoly on service

Gartner Group Software Maintenance "Bill of Rights" Concepts

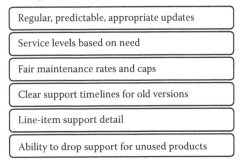

FIGURE 14.2
Software maintenance code of conduct.

price. This applies to hardware vendors as well as software vendors in any situation where there is no competition. Many users have attempted to escape one vendor for a competitor, only to see the competitor purchased or driven out of business. For this reason, many buyers are attracted to open systems products and SaaS products where the relationships can be better controlled.

Just as with information technology (IT) hardware, there are no mandated fixed support timelines for software. The OEM (developer) is in total control of the period of time they wish to support any of their products. Users are therefore vulnerable to changes in support cycles, but the lack of regulation in this area is also a negotiating opportunity. Developers cannot be forced to support older products if they do not wish to do so. Users should ask legal counsel to help structure negotiations at the outset to control source code if the developer abandons support.

Lack of line-item support pricing is pervasive with software pricing, including pricing for vaguely described "Machine Code" and "Embedded Code." Users have to take control of this issue by requiring, as a condition of purchase, that such clarity be provided. Hardware is separate from software for financial accounting, so any overlaps should be exploited in requests. Software that is licensed to specific hardware may be more easily separated, if only for the need to track the software "Stack" as part of an asset management system.

Among the most vexing issues for software buyers has been the inability to remove unused licenses from maintenance contracts. The problem is brought about by license agreements that are volume based, such as *per seat* or *per user* licenses. If the number of *seats* falls below the agreed threshold, the licensor is not willing to provide a refund or rebate without rewriting the original contract as though the additional seats never existed. Because there is no competition for support outside of the developer, the user is stuck with whatever terms it struck at a different point in time.

Jailbreaking

Jailbreaking has been in the news for several years, initially as a cell phone issue. Jailbreaking came about as users of the Apple iPhone chaffed at the limitations built into the iPhone limiting purchases of apps to those offered by the App Store. Users created code that unlocked the phone, figuratively taking it out of "jail," and accessed and used apps not sold through Apple. The Electronic Frontier Foundation (EFF) pursued the matter with the

U.S. Copyright Office on behalf of consumers asserting that the purchaser of the phone had the right to manage the use of the product postpurchase. The U.S. Copyright Office agreed, and jailbreaking of Apple phones was made legal.[2] Unfortunately, the U.S. Copyright Office ruling only applied to Apple brand cell phones and not to any other devices.

Technically, jailbreaking requires accessing and then modifying the machine code layer to facilitate using functions not originally provided by the manufacturer. In the case of Apple, the code is called the IOS (internal operating system) but the name is not the delimiter. All machines come with machine code and it is this layer of code that is used to control very basic functions, such as authentication of attachments or higher-level software (apps). Other wording for a similar function is accessing the boot or root code, bootloader, firmware, or basic internal operating system (BIOS).

The cell phone *unlocking* function is also controlled at the machine code level. One set of code changes is intended to allow access to apps sold outside the App Store (jailbreaking) and a different set of code changes modifies the authentication of the cell phone to use a different home network.

Unfortunately for the vendors of media and content, the same skills used to write unlocking and jailbreaking routines can also be used to modify machine code to circumvent protections built into machines that enforce such things as game copyrights or ink cartridge vendors. As a result, there is a strong lobby coming from the "Content" industry against allowing jailbreaking or unlocking for other devices, as they see their particular financial interest in selling copyrighted games could be jeopardized.

Both unlocking and jailbreaking are illegal under U.S. copyright law, unless an exemption is granted by the Library of Congress. (The Library of Congress has jurisdiction because the machine code is being claimed as software and thus falls under copyright law.) It has been widely argued that the restrictions on end users that flow from such technical legal limitations is not appropriate to life in the 21st century.

There are changes being contemplated in the U.S. Congress, with the support of the White House, to make unlocking legal, but the method chosen may not directly open machine code to users in general. Unlocking may begin the discussion about software locks and the rights of buyers to control their whole machine but it is not going to be legalized in the near term for any items other than cell phones. Unlocking or jailbreaking in very general terms are the purview of federal copyright reform, which may never happen or may happen badly. Users still need to protect their own interests through education and negotiation.

Support of jailbroken/unlocked phones is and has been a contentious issue. The vendor of the cell phone may refuse to repair jailbroken phones under warranty. This is technically absurd. Repair of a broken connection has nothing to do with the apps being downloaded. The vendor is simply using the threat of withholding repair to prevent users from taking advantage of non-OEM provided services.

On the other hand, withdrawal of support of the operating software (IOS or machine code) is legitimate in that the developer of the machine code cannot diagnose problems and provide fixes to code that they did not write. Users of jailbroken machines need to understand that in order to get support for the machine code they will have to restore the original version. Owners need to understand how they will restore a supportable code set before jailbreaking or face the possibility of zero support.

It remains possible that the U.S. Congress and European Union commissions will reform copyright law so that users will have broader rights to control the use of their purchased equipment. However, reform may be years away, and bad laws are often passed in the name of "reform." In the meantime, consumers of all sizes are able to protect themselves from such limitations through the power of the purchase order.

APPLICATION SYSTEM HARDWARE DEPENDENCIES

Application developers write products based on a selected operating system. The operating system provides many of the common routines needed by a complex program, which would otherwise need to be written out repeatedly. These commonly executed groups of instructions may be called subroutines, macros, and more recently, APIs (application programming interfaces).

Operating systems are themselves developed using the instruction set provided by the hardware manufacturer. The available instructions are fixed. If a new instruction is needed, only the hardware manufacturer can create it and add it to a new model of hardware.

Consider the analogy of buying building toys for children. Legos and Lincoln Logs are building toys, both have interchangeable parts, but the formats for the parts are different. The shapes of available parts represent the different instructions available in the operating system. As with older models of computers and older operating systems, both manufacturers of building toys began with a small set of interlocking parts and as the products became popular, more shapes and colors were added. The original format of the design and design parameters remain intact. As a result, all older parts still work with newer parts in the same general format. New types of blocks are added to the kits in much the same way as operating systems are enhanced to supply new hardware features.

Each type of building block format has advantages and limitations over the other. Before buying the building kit (the application), the purchaser has to decide on the functionality of the final product. There is no substitution between Legos and Lincoln Logs. Buildings constructed using the different formats may have similar features, such as doors and windows.

The analogy works because machines are different on the inside even if the results have common appearances. Emulators and virtualization make it possible for completely separate types of programming to share the same machine, but they remain separate.

One of the only ways to avoid hardware dependencies of software is to use applications provided as a service, without regard to the machine. This does not eliminate instruction-level limitations; it puts them elsewhere for someone else to worry about.

Configuration-Dependent Licensing Models

Operating systems are assumed to be always in use at some base level, so licensing is generally keyed to the speed of the processor or quantity of processers. If there is a volume-variable license rate for the OS or systems software, it is tied to the usage of the processor. Usage-based models are validated by performance monitoring functions of the operating system, which have been common on systems for decades.[3]

In server and enterprise settings, the more powerful the central processing unit (CPU), the higher the license and software maintenance costs. With enterprise-scale processors, the licenses are usually variable by the quantity of licensed processors, since many machines are preloaded with the full complement of processors, some of which can sit dormant until activated. The activation of these additional processors can be done remotely once the billing arrangements for the additional fees are made.

Hardware designs that incorporate multiprocessors and multicore processors are challenging from a license pricing perspective. Many users discover late in the purchase process that their choice of a new processor has catapulted them into a new category of license fee for some element of their software stack. Even if the basic OS may be favorably licensed, third-party systems software products do not have to follow suit.

Software maintenance of all OS and systems software products is always the domain of the provider. There are no legal options to acquire software service of proprietary products. Open domain versions of some old operating systems are available and can be serviced independently. Open systems/open domain products such as Linux are extremely attractive as alternatives to proprietary OS primarily to control the ongoing costs of software support.

Usage-Based Licensing Models

Application systems licenses are commonly priced according to the number of users (seats) and not just a single license covering all instances. Variations include licensing the number of concurrent users, such as with products used intermittently across a network or enterprise-wide licenses where a particularly large end user can offer a consistent block of revenue in exchange for less license management.

Identifying and validating all the instances of the *per seat* or *per instance* contract is far more complicated than for licenses by serial number or even when tied to processing usage. Application vendors offering these options are wary about enforcement. It is common in these types of license agreements for vendors to require users to allow access to records of usage, login credentials, and any other tools that might be available to help validate usage. These same vendors are usually more aggressive in requiring audits. This is not surprising since the history of such licensing models has proven high levels of license misuse on the part of users.

Maintenance and defect support for applications is the unique responsibility of the product developer. There is nothing done at the repair level that impacts applications, as the hardware must operate before the operating system and the OS operates correctly before the application.

"BLACK BOX"-TYPE PRODUCTS

The "Black Box" category of product appears as a hardware item but includes significant software that is not separately licensed. The difficulty with these products is that repair is difficult to separate from the software, often to the extent that repair is unavailable except through the OEM. Since the proportion of hardware and software is unknown, it is unclear from most *black box* vendors just how much, or how little, ownership is really possible. Accounting for these purchases can be tricky as some equipment can be transferred, in which case the product can be accounted for as a capital expense, or a transfer involves relicensing, in which case the equipment is not really tangible and should not be depreciated.

An example from the automobile industry helps explain the challenges facing buyers. Today's automobiles are effectively IT appliances in that they include many electronics essential to the operation of a car, from engine timing software to ABS brakes to keyless entry systems. These electronics include powerful microcode and firmware, all of which are currently expected to transfer with the vehicle. Buyers of automobiles do not separately license this software. In contrast, while many vehicles come equipped with satellite radio, the contract is offered as a separate service.

The sticky area is the ability of manufacturers to restrict access to their dealer network for reloading or updating of the firmware, which would intrude upon the rights of the auto owner to seek service independently. The potential for OEM control of end-user repair and transferability through manipulations of microcode and firmware remains large and is a risk in all industries.

VIRTUALIZATION

Virtualization is an expansion of the concept of device emulation. IBM began selling its Virtual Machine (VM) architecture in 1982, allowing multiple imaginary machines to run concurrently at the same time. The physical memory and processor resources of the machine were managed by the VM software so that each machine could operate separately. The quantity and variety of operating systems that could be run simultaneously was initially limited but capacities grew and software was improved. Today's single physical machines can run thousands of virtual (imaginary) machines simultaneously.

Emulation was also widely used since the early 1980s to allow "non-native" devices to attach to a channel. This eliminated the need for the processor vendor, such as IBM, to write support for a competitive product in native mode. For example, the early line of EMC 4400-4480 disk arrays was built to appear to the host system as a 3390 disk drive. Although EMC could have built plug compatible disk drives (as has been done by StorageTek and Hitachi), the ability to attach via emulation was key to the disk array innovation that now dominates the storage industry.

Virtualization is sold as a cost-saving tool allowing what may have been hundreds or even thousands of small individual servers to be consolidated into a single, centrally controlled machine. The end user does not see the change because the entire configuration including all the software is emulated through the virtualization software. The ratio of compaction possible using virtualization varies dramatically, but if many of the smaller servers were operating below capacity, the process can be extremely cost-effective.

Machines made to support virtualization are physically more complicated, but there are far fewer of them. The largest problem with virtualization from a hardware support perspective is that one machine failure can (and does) take down hundreds or thousands of applications running at the same time. The impact of a small interruption can be large, whereas if a single server were to fail, it might only inconvenience a single department in a remote location. Virtualization therefore is best done on highly reliable equipment in secure, climate-controlled, and power-conditioned settings.

Software support is far more complicated in a virtualized system. Not only is there an application system to support, but the operating system running underneath the virtualization software must be supported and then the virtualization software itself must be supported. Virtualization also offers the opportunity to run operating systems and applications not created by the OEM, which really complicates the support picture.

One of the unusual support problems brought about by virtualization is the confusion on the part of the desk/service desk regarding which system has failed. The end user no longer knows the serial number of the enterprise server now hosting the virtual system, so can only reference the original server name. If the help desk/service desk does not have a complete list of which virtual servers reside on which physical servers, connecting the support request with the appropriate support team can be time consuming. It is also common for internal asset management systems to continue to list servers that had been physical assets, now virtualized, as assets. This often causes downstream problems with removal of ghost assets and the associated costs of support, such as insuring servers that are virtualized, or paying for licenses or support contracts for other assets that have been removed due to virtualization.

Software pricing for virtualized machines is not regulated. If the developer of the licensed product wishes to charge preposterous prices for moving the license from one small server to a virtualized server, they can do so. The user of licenses that do not transfer at a reasonable cost can keep the old machine running or replace the application. The only assurance that any user has of being treated fairly is the presence of competition.

CLOUD AND SOFTWARE AS A SERVICE

The cloud is a marvel of modern engineering and marketing. Engineering to use the Internet for connecting millions of disparate end users and data points is brilliant. Marketing is equally brilliant to give consumers the illusion that they are no longer using hardware or software licenses.

Every bit of data stored in the cloud is stored on hardware somewhere in the world. Backups of data collected in the cloud are also stored on hardware (or digital media) somewhere in the world. Software programs are executed on processors somewhere in the world where the support issues of real hardware and licensed software still connect. A major cloud host is a gigantic data center. Cloud data centers tend to be located in areas where there is a stable power grid, with widespread access to networks at the highest possible speeds. In addition to the skilled labor needed to attend to the constant physical needs of hardware break–fix, the labor force for the software management of the cloud can be located anywhere in the world.

Cloud data centers are therefore no less vulnerable to software or hardware failure than their corporate or government kin. Contracts may shift from the name of the corporate owner to the name of the cloud host, but the contracts do exist. As with any other forms of agreement, it is important to pay attention to how to exit a cloud agreement before starting. Cloud vendors can be excellent or awful, solidly financed, or flimsy. Equipment selections can be stable or unstable, and software licenses may be incomplete and subject to audit. Contracts that anticipate bringing operations back in-house or shifting between cloud providers are worth the time to create on the chance that all is not perfect in cloud paradise.

An extension of the cloud concept is Software as a Service (SaaS), which provides a specific application through the Internet without any specific data center equipment or licenses. SaaS products remove the burden of selecting and maintaining different software versions for different and possibly incompatible platforms. Well-crafted SaaS solutions offer instant availability and compatibility with nearly any device with an Internet connection. (There are always some caveats and limitations on connectivity.)

The downside is that the SaaS application cannot be customized in the same manner as a licensed product. Although the product may be flexible in terms of field labels, the underlying operation is fixed for all users. Users selecting SaaS instead of licensed applications are making a trade-off between instant gratification and customization.

The SaaS vendor also completely controls the hardware platform and the associated support services. This may be a factor in response time through the Internet, vulnerabilities to disasters, lack of adequate backup, slow service restoration times, and no control whatsoever over the hosting environment or location.

SUMMARY

Software support terms and conditions are tightly controlled by the OEM or developer. Controlling the maintenance costs of software licenses requires careful attention to the future impact of terms and conditions in the license agreement. It is not adequate to negotiate the first few years of a warranty and hope for the best beyond the warranty. The most costly part of the long-term use of software is the long-term cost of maintenance, and those who prepare for 10 or more years of use will not regret the value of planning ahead.

NOTES

1. For more on the attempt, see "Gartner Global IT Council for IT Maintenance Develops Code of Conduct to Address IT Maintenance Consumers' Most Serious Concerns," July 19, 2010, http://www.gartner.com/newsroom/id/1403313.
2. For further discussion of the broader context of jailbreaking, see "IOS Jailbreaking," *Wikipedia*, http://en.wikipedia.org/wiki/IOS_jailbreaking.
3. CPU usage monitoring has been common for operations managers as a key to making sure that loads on the system are in balance and service to users is within tolerances. In addition to manufacturer supplied tools, there are a wide variety of independent software tools that purport to measure various types of utilization and alert managers if thresholds are exceeded.

15

Support Restrictions to Control Purchases

INTRODUCTION

Manufacturers have long used pricing for support and maintenance to guide customers into new product purchases. The goal of the manufacturer is not to help you keep your old equipment in service but to entice you to buy a replacement product. Many policies promoted by original equipment manufacturers (OEMs) are designed to create the fiction of urgency, particularly when the stockholders of the OEM are commanding new revenues.

ADVANTAGES FOR VENDORS

The increasingly monopolistic terms and conditions offered by vendors are the result of a long history of selling small contractual inconveniences traded for price concessions. It has been no surprise to those of us in the industry in the 1980s and 1990s to predict that vendors would continue to rearchitect their products to create "mini" monopolies for themselves. Earning pressures have not abated. Projections for hardware margins continue to decline so new sources of revenue (and margin) have to be developed. It is obvious that selling services and support to existing clients is easier than selling more products at depressed margins.

This is good for the vendor but bad for the user. The user should never fail to negotiate the best possible terms and conditions for their enterprise assuming that profit margins will simply shift elsewhere. Margins for support and maintenance have remained extremely high even as other margins are winnowed by competition. Allowing vendors to monopolize support and maintenance simply shifts user profits to vendors without adding value.

Profit pressures exist in every industry throughout history. Buyers who accept that the information technology (IT) industry is uniquely deserving of retaining high margins are not supporting their employers who also face profit pressures for their own product sales. Even government buyers have a fiduciary responsibility to the public to be circumspect with public money.

Buyers of all forms of technology products are now faced with reasserting their power as buyers to demand appropriate terms and conditions. Each new negative policy change that is tolerated in the marketplace is later adopted by competitors, in much the same way that baggage fees became acceptable to airline passengers. Fortunately for technology buyers, the opportunity to use the power of the purchase order to dictate more favorable terms can be used more effectively than a single ticket purchaser trying to avoid paying baggage fee.

ACCOUNT CONTROL

Technology vendors use an insider phrase, "Account Control," with true intention in the technology industry. Once a client has committed to a platform, the vendor will work the account to assure continued and expanded investments in the product line(s), expanded use of the vendor-provided "service" offerings, and nearly monopolistic control over software support and equipment repair. Many technology purchases thus evolve into a marriage without benefit of prenuptial agreements. As with any marriage, rules are difficult to change after the wedding, so the time to prepare is before walking down the aisle.

Most managers involved in equipment or platform selection focus first on meeting the technological needs of the application, then price, and then turn over the contracting to a procurement or legal team. This process does not deal with the reality that the vendor intends to develop an entangling relationship over time. The opportunity to avoid future pain and suffering over an unhappy vendor relationship (hardware or software) requires that the entire potential period of use be considered, including rarely considered topics such as:

- How to keep products up and running beyond the warranty offerings of the vendor
- How to support products that are beyond End of Service Life (hardware) and End of Support (software)
- How to resell, trade in, or dispose of surplus assets, licenses, and support agreements
- How to avoid being trapped by the planned obsolescence objectives of the vendor

The focus of this section is to provide the background needed to seriously negotiate on a level playing field with vendors of hardware, software, and associated maintenance and support services.

Most readers will find that the vendors themselves, particularly the local representatives, are unaware of why policies exist, and many will not be able to parry questions without further research. This is a good thing, since the more legitimate challenge one can bring to a product selection or negotiation, the more seriously the vendor will take the discussion.

This section discusses strategies for improving negotiations and selection of appropriate break–fix contracts based on knowledge of the need for service, the goals of the organization, and understanding vendor marketing tactics that are used to prevent need-based negotiations.

Wolves in the Hen House

The problem of excessive vendor control starts with allowing vendors to participate in the project planning. It is axiomatic that "he who writes the RFP (request for proposal) wins the business." This is not to say that vendors should not be invited to offer recommendations and use their expertise and imagination to bring to bear on business problems but rather to warn that users are going to be taken advantage of when the relationship is out of balance.

The longer the vendor has been incumbent in the account, the more likely that executive level relationships have been forged high above the selection level. This includes boards of directors and the C-Set (CIO, CFO, CTO et al.) plus alumni relationships, country club relationships, and networks of former colleagues and professional associations. These are powerful bonds that are used by executives without the involvement of the sales representative. Many otherwise logical selections have been undone or blocked by a single phone call.

End users and vendors are never really partners. Vendors have to make money to stay in business, and end users want to pay as little as possible for the same purpose. This is the natural relationship and is always one of constant tension. Arrangements that seem otherwise are almost certainly unfavorable to one party or the other.

Vendors use a common variety of sales tactics to persuade users to enter into contracts that are more favorable to them than the end user suspects. Users can reject or negate these tactics. The first step in preventing imbalance is to be aware of the tactics and plan ahead for response.

Tactic: Controlling the Request for Proposal and Award

The ideal position for a vendor is to be in control of the specifications in the request for proposal (RFP). This is typically done by the end user who has worked closely with a favored vendor and already negotiated the equipment selection and does not want other bids to be competitive. Even in situations where the procurement department exercises its authority and solicits competitive bids, the bids are effectively rigged to guide the purchase along preordained lines.

Often, awards are made informally and agreed "verbally" pending the official purchasing process. Savvy vendors, particularly incumbent vendors, strong-arm the management team outside of procurement to insist that a verbal agreement exists and must be honored. Since the procurement team is often told they are insufficiently technologically savvy, the verbal purchase order turns into writing.

Tactic: Monolithic Purchases

Multiyear contracts are often awarded to a single vendor for ease of administration and consistency of product. This is common in situations where the identical product needs to be deployed throughout the enterprise to control support costs. Problems with this strategy and digital products include:

- The OEM may not be able to provide the identical suite of components across multiple manufacturing runs or over time, thus defeating the original intent, other than for price.
- Identical components of the same age across large numbers of devices, if flawed, expose the owner to the risk of widespread concurrent component failure.

- Maintenance activities in the future may require substitution of different components or models, thus defeating the value of support for identical models.
- The OEM will have a multiyear exclusive agreement that does not have any competitive elements. In these situations, most OEMs become complacent and service attention declines even if replacement product pricing is indexed in some competitive manner.

Monolithic purchases are also challenging to administer for purposes of version-level tracking and patch management. Although many devices can be updated remotely, there are risks to remote dispatching of new sets of updated code. It is often the case that very large and widespread outages have resulted from incompletely tested or poorly planned remotely dispatched "updates" that instead crashed the local system. Large deployments of nearly identical products are at high risk for large-scale updates based on identical configurations that may prove to be more variable than planned.

Tactic: Strategic Partnerships

Treating technology vendors as "Strategic Partners" is particularly dangerous. There are very few instances outside of a formal joint venture where any vendor is ever really a partner. (It is axiomatic that if there is no money at risk in the partnership game, such as an equity stake, there is no partnership.) This is not to be confused with a product (usually software) custom development project where the vendor is deeply integrated into the development team and employees may be shared.

The strategic partnership for commercially available products is the best possible deal for the vendor who gives up little (usually a lower price point today) in exchange for a set of golden handcuffs with which to bind the buyer into future purchases. The user gives up all control over price and quality in exchange for having reduced the number of vendors vying for the business and an ostensibly lower price point.

Tactic: Trimming the Vendor List

Reducing the number of vendors for true commodity items (carrots, toilet paper, pencils) may be useful to the accounting side of the business but is a step backward for technology purchases. There are three reasons:

1. "Commodity" technology products are built using commercially available components rather than proprietary components. A Dell, HP, Gateway, Acer, or Lenovo personal computer is considered a commodity item, but the ergonomics and internals are quite different. One can run an "open" design Windows or Linux operating system on any of the aforementioned, but the products are far from identical.
2. Regardless of the proprietary or *open* nature of the components or the software, the OEM is in charge of the pricing even when the product is sold through channel partners or distribution. If one wants to cut down on the number of vendors, one can cut all but one representative of any particular OEM and still have competition between OEMs.
3. Component manufacturing and finished product assembly is largely done in Asia by specialist companies, many of whom produce products for multiple OEMs in the same line. Reducing the number of vendors does not by itself assure any changes in the quality of products being produced, which is the real issue for users. If all OEMs that use the low-quality assembler are removed from competition, the result will be lower quality for all purchases.

Tactic: License Audits and Compliance

Software vendors are acutely aware that their products can be pilfered, replicated, and run illegally. Most terms and conditions give rights to audit license compliance to the vendor, which many use to their advantage. As with everything else, some vendors are more honorable and flexible than others when it comes to compliance issues.

Owners are occasionally faced with threats of license audits in order to frighten the owner into some other agreement, not even necessarily having to do with licenses. If an OEM were to demand a hardware service contract on all machines in the enterprise or else, the "or else" is a hint that a license audit would be forthcoming. There are vendors known to be extremely aggressive in license compliance, which is their right. The question for end users is how to react other than by rolling over. Corporate counsel should be involved anytime such threats are made, as some threats are hollower than others.

Tactic: Return to Service Premium

Unlike hardware maintenance, which is event driven, software support is product driven. Flaws in each specific version of the product impact all users, so all maintenance work (problem diagnosis and correction) is widely applicable. Users that drop support for a period of time and then want to return to service have not paid for support in the interim. If that support includes defect support updates and fixes to routines that they would have used, they would be free-riding on the support contracts paid for by others.

It is therefore reasonable to be charged for return to service for software support if such support included recent work for defect support. If the product has been static for some years, it may not be reasonable to be charged, which is a choice of the part of the user. Upgrading to a new version is almost always a chargeable event and usually involves some price negotiation for the period of lapsed service.

This is a negotiating point for software support contracts and not an excuse for vendors to make similar demands for hardware support contracts.

PLANNED OBSOLESCENCE

Users need to start their product selection and negotiation with the expectation that the vendor wants to keep selling new versions of their product as quickly as possible. It is in the best interest of the vendor to create events that lead the buyer to replace the product sooner rather than later. This is called "Planned Obsolescence" and is so common in technology products that it is not even noticed.

This type of planned obsolescence is not the same as engineering the product to fail the moment the item is out of warranty. Electronics are too fragile to be manufactured to intentionally fail quickly, because the consequences of supporting a massive in-warranty failure would be overwhelming. This statement does not mean that some electronics fail at high rates, only that there are few incentives for manufacturers to deliberately select components to fail.

Tactic: Require New Versions to Access Routine Functions

QuickBooks is a popular accounting package used by small business. One of its most powerful functions is to access tax tables for sales tax and employee payrolls. In addition to the tax updates, accounting does not really change, so older versions of the software would work indefinitely.

In order to generate new product sales, QuickBooks deliberately restricts access to the tax tables to those with the most current version of their software, forcing buyers to repeatedly repurchase the entire package.

Tactic: Manipulate End of Support and End of Service Life

Many software vendors announce End of Support (EOS) as a threat to their clients to pay for upgrades to new versions or to purchase indefinite "maintenance" agreements. Hardware vendors do much the same thing by announcing End of Service Life (EOSL) to prod happy users to replace older equipment and then further clamp down on access to support routines (such as onboard diagnostics) to prevent competitors from offering alternative solutions.

Tactic: Refuse to Allow Transfer of Equipment, or Prepaid Support Contracts to Secondary Buyers

There are vendors who will not permit an owner to transfer the equipment they purchased to a different location, or subsidiary, or used equipment buyer without permission. This dramatically degrades the value of the equipment and may even interfere with the ability of the owner to get financing for the purchase. On the topic of location restrictions, the vendor will state that they might not have support teams available at the location. This is a spurious argument since equipment vendors can easily contract readily for independent support providers at remote locations. One might also question why the product would be so difficult to repair that well-trained technicians are incapable of repair.

In the case of refusals to transfer hardware equipment to subsidiaries, vendors are on very weak ground. The most common excuse provided is that the subsidiary is a separate corporate entity that lacks independent financials and may not have a credit history as strong as the parent. Although this may be true, the purchase has already been made. The vendor has been paid and their only possible collections risk is for payment of support agreements. Even if there is a lessor involved, the lessor has taken on the credit risk and thus can approve or decline the transfer of responsibility for payment, but cannot refuse to allow changes of location within the United States.

There are rational reasons for keeping equipment in the country where it was originally sold but at the option of the owner, not the OEM. The owner may lose warranty coverage when the equipment is moved across borders, as the OEM will have likely set up its warranty coverage to be country specific.

Transfer of prepaid support agreements or extended warranty should not be at the discretion of the OEM, but if the contracts were written with these restrictions the deal is the deal. Just as OEMs cannot go back in time to modify contracts, neither can users.

Tactic: Restricted License Transfer, or Software Maintenance Contract within the Enterprise

There are no technical reasons why licenses or associated maintenance contracts cannot be transferred easily between parties, including within an enterprise. Unlike hardware, where support may mean a physical logistical challenge, software can be supported anywhere in the world without any physical costs. Transferring a license from one serial number or location to another is a wholly administrative process. If manufacturers claim that the process is costly, it is only because they choose to be inept. Matching serial numbers in an entitlement database is a common application for manufacturers of all kinds, from bicycles to whirlpool bathtubs.

Most licensed software tech support is delivered using a worldwide network of software staffers often selected for their availability around the clock. Many domestic and overseas support call centers provide back office software support for multiple vendors, fully detaching the support of licenses from the possibly unique skills of the vendor's own staff. Problems that are not resolvable by the support vendor are escalated to higher-skilled employees and eventually to the vendor if the problem is fully identified. The vast bulk of calls for support (roughly 90%) are for problems that are associated with settings, authorizations, and routine operational questions rather than for actual problems.

Refusals to transfer licenses must therefore be based on some other objective. The most likely culprit behind such refusals is the opportunity to extract new license fees from a user that has no bargaining power in the negotiation.

Tactic: Refusal to Transfer Licenses or Maintenance Contracts to a Secondary Market Buyer

There is no one policy for license transfers of machine code, operating systems, or application licenses. There is no consistency of law domestically or internationally. Negotiations on this point are both possible and essential.

For example, Microsoft used to license Windows to the user and not the machine. It was an unwieldy process and caused most business users an enormous amount of trouble in tracking physical license documentation. This must have been a repeated point of contention between large users and Microsoft, as Microsoft eventually changed its licensing to flow with the machine. Machines now sport stickers showing the license that now travels with the serial number.

Although the Windows operating system license travels with the machine, the application layer, such as Microsoft Office, still travels with the user. This facilitates the transfer of older machines to secondary buyers and allows the original owner to use the Office license on a new machine.

Unfortunately, many manufacturers try to avoid allowing a prepaid service contract to pass between buyers. This is analogous to an auto manufacturer refusing to transfer the warranty between the original owner and the secondary buyer. Most car owners would reject any purchase with such restrictions, but technology buyers have allowed themselves to be pushed around by vendors to accept vastly different terms and conditions.

Buyers, particularly large buyers, have the influence to insist on better terms, particularly the right to transfer any remaining warranty or prepaid support contract, to a secondary buyer. If this is neglected, the value of the asset is severely diminished.

Tactic: Refusal to Allow a Used Market for Licensed Products

The European Court of Justice in the European Union, the equivalent to the U.S. Supreme Court, ruled in 2012 in a case brought by Oracle against UsedSoft GmbH, that license holders can resell their perpetual (purchased) licenses freely in the secondary market provided that they no longer use the product.[1] This is expected to change the way software is developed in the European Union and elsewhere as vendors experiment with ways to manage their margin expectations in a marketplace where older versions may be competitive with newer ones. It may be impossible to fully assess the impact of the policy change as vendors are also shifting their marketing toward SaaS (Software as a Service) and cloud offerings.

FEAR, UNCERTAINTY, AND DOUBT

Vendors are enthusiastic users of marketing techniques known as "FUD" (Fear, Uncertainty, and Doubt). Without FUD, many vendors would be unable to persuade their users that they must replace perfectly well functioning products with the latest version. Figure 15.1 shows common examples of FUD.

Without the Vendor Maintenance Agreements, the Equipment Will Fail at Higher Rates

This assertion is logically flawed on several levels. First, the circumstances under which equipment breaks is a function of design and manufacturing. The rate at which equipment can be expected to need repair is a constant and not related to the logo on the technician's nametag.

Second, it implies that the technician dispatched from the OEM is competent, whereas the technician from any other source is incompetent. This is

FIGURE 15.1
Fear factors.

also a false assumption. In many cases, the pool of qualified technicians is shared among many OEMs and independent providers alike. There are businesses supporting the labor needs of both including tracking of certifications, dispatching, and billing. If an OEM claims unique service technicians, it should cause eyebrows to be raised. A simple request for a written guarantee that all technicians are direct W2 employees from a company executive will likely end the discussion.

The most legitimate claim to better service by the OEM is with respect to parts for new equipment. OEMs control their own parts desk, and they can choose to be either lazy or efficient in filling parts orders from outside their technician network. Users can demand priority access to the parts desk in their service contracting agreements to bridge the gap between the time when the product is new in the market and the time when service parts can be acquired through alternative sources such as distributors, surplus inventory, and used products.

More Costly Service Contracts Result in Better Service

This myth is actually two poor assumptions in one. First, the relationship between service quality and price is not well established. Vendors react to complaints, not hugs. Often, the satisfied user gets little attention regardless of premiums paid. Second, in the case of equipment failure, not needing service is far preferable to needing service. So the payment of premium pricing for purposes of extra attention is not correlated to equipment failure. The Service Level Agreement (SLA) governs how quickly the OEM is obligated to arrive to make a repair, which is a factor in service restoration, but not the only criteria for "better" service. The best service would be the least consistent mean time to repair, not the mean time to arrival.

The "insurance" value of such arrangements is also questionable. Users with the highest-level SLAs can still be caught short in a major regional event, such as a hurricane or flood. Under these circumstances all customers with high-priority contracts will be equally difficult to support and technician resources will be thin or unavailable for the same reasons as the general emergency. In these cases, the premium-priced SLA has no value, and repair and service issues should be included in a disaster plan.

The vast majority of service incidents are unpredictable and not linked to time or geography. If any OEM is unable to handle a T&M (time and materials) call or low-level call at the same time as a high-priority call, then its technician resources are entirely too thin. Under situations of technician scarcity, even the buyer of a premium plan will have no better support than those with a bare-bones plan.

Although technician scarcity seems a legitimate concern, the use of on-call and "flexible" labor for service delivery on the part of OEMs all but prevents this specific concern. Although there may be some OEMs with very specialized equipment and limited resources, the vast bulk of technology assets are not in this category. Much as an alternator in a Ford is not dramatically different from one in a Chevy, a memory card in an HP server is not dramatically different from a memory card in an IBM server.

Fear of Complexity

Hardware and software have to operate in concert to execute applications, but that does not mean that they are a single product. They are not intertwined except in contracts. The same is true of claims that products are legitimately complex technically, but the hardware and licensed software elements are still separate "complex" combinations of technology hardware and software, which are also separate.

There is no actual linkage between the support of licensed software (software maintenance) and the break–fix process (hardware maintenance). These two types of support are linked only at the marketing level, and as such can be *delinked* as part of a contract negotiation. Vendors that make the intertwined claim are doing so in order to cause clients to purchase both hardware and software support agreements together.

The myth of complexity only works if the end user accepts the idea that patches to the operating system (licensed or open) makes corrections to hardware (physical) problems. This is not true and has never been true. Products that need a physical repair (such as a broken wire) are not the same as products that have an instruction bug. If the instruction bug is at the machine code level, then the patching must come from the hardware engineers. This is the case for proprietary, appliance, and commodity hardware.

The financial reporting structure of the OEM almost always separates the hardware support (maintenance) revenue from software maintenance revenue. Employees providing support for hardware or software are typically reporting to separate management and often through separate profit centers. This is a necessity as repair remains a world of "warm hands" requiring the physical presence of a technician who is subject to labor laws, withholding, and typically is backed by bonding or insurance. In stark contrast, the software technician can be located anywhere in the world and working under any conditions.

Further evidence that hardware and software support are entirely separate functions is found in the call triage function of the initial trouble call. When calls are placed for service or support, the series of questions posed by the help desk or service desk are designed to quickly differentiate the problem to be assigned to the appropriate hardware or software service specialist. Claims of *complexity* do not survive the initial triage steps in opening a support ticket.

Although the aforementioned minimizes the challenges for supporting both hardware and software, the process is the same for all products. Efforts to confuse the two are deliberate on the part of the manufacturer seeking high-revenue service contracts. Marketing representatives for the OEM are not likely to fully understand the differentiation, so be prepared to escalate the discussion to management.

Multivendor Conflicts

It is often the case that not all products work well together. This includes products within a product line as well as competitive offerings. The reason is that a large swath of products are designed and built under contract, usually with an Asian provider, to meet a certain market need. Compatibility between products made in this arrangement is not a high priority and conflicts can be expected. Designers of proprietary products have more control and should be able to keep products compatible, but there is room for error in all endeavors.

There are vendors that claim they alone can provide a fully coordinated suite of equipment and software. The inference is that it would be a terrible error to open any element of the IT project to other vendors. These types of claims should be treated with some skepticism, as there is a long and successful history of competitive products built intentionally to be used as "plug compatible" alternatives.

Software product compatibility is even more difficult to achieve particularly when software vendors grow through acquisition. Many different products become sold as a "package" when in fact the real connection between them is the marketing materials. Buying the complete package is no guarantee of compatibility, only the comfort of having a single phone number to dial to request support.

There are vendors that offer comprehensive multivendor maintenance agreements covering all the products at a location (commonly within a data center). OEMs that engage in these offerings subcontract with independent technicians to handle products they do not themselves support. Vendors do not like to hire competitors as subcontractors if they can avoid it. This is the same as hiring a general contractor to remodel your kitchen, paying to have a single point of contact rather than hiring each trade separately. There is always a cost to hiring the general contractor, which has to be weighed against the value provided.

Hardware FUD

Manufacturers prefer to sell more hardware more quickly and churn users quickly into replacing older equipment. Buyers would prefer to keep equipment installed as along as possible. This is a natural tension.

Legitimate improvements in technology provide some enticement to buy new equipment, however, most businesses do not need the latest and greatest of anything. The impetus to replace equipment is often artificially stimulated by the OEM with a series of marketing tactics playing on fear that older equipment is dangerous to keep in service. Thus, FUD is a major driving force behind new equipment sales.

There are countless examples of products that are sold with a "standard" 3-year warranty (itself a fiction). Pricing for continued service at the expiration of the warranty is made intentionally unattractive compared with new equipment. Some manufacturers add insult to injury by refusing to allow continued access to the machine code, which is otherwise essential to keeping equipment in service. Between the two policies, most buyers are subtly compelled, if one regards a hammer as subtle, to constantly replace equipment on intervals dictated by vendors and not need.

At the expiration of the manufacturer-provided lease or warranty period, users are directed, due to FUD, to pay whatever the manufacturer commands or seek competitive equipment replacement. If FUD is ignored, alternatives for independent support are only viable if the owner has control of essential machine code. Competition for hardware break–fix support agreements will effectively justify the long-term use of fully depreciated equipment without the unrealistic threat of competitive hardware replacement.

Physical FUD

Manufacturers also strong-arm owners into replacement cycles of their choosing by making claims that the equipment will not survive without their support agreement. These various claims are so endemic to the perception of manufacturer support that they are accepted as true, when in fact all these assertions are false or intentionally misleading.

- Equipment will quickly be physically obsolete and should be quickly replaced.
- The OEM will drop *defect support.*
- OEM technicians are superior and uniquely qualified for complex tasks.
- OEMs can guarantee noncounterfeit parts.

Physical Obsolescence

Physical obsolescence is rarely observed in electronics that are repaired as needed. Most equipment in climate-controlled settings have failure rates that are the result of manufacturing and design, and not the degradation of materials. With the exception of capacitors, most electronics kept out of wet and hot locations, and buffered by surge protectors continue to function far longer than most applications are in use. Moreover, manufacturers do not actually know the complete useful life of the products they create because they cannot track the equipment repair requirements beyond the warranty period. Whatever assertions are made about physical useful life are notably unsupported.

Fear of Losing Defect Support

Loss of defect support (also known as patches and fixes) is a powerful driver that pushes owners into exclusive postwarranty support agreements for both hardware and software. In order to control the reaction to the vendor FUD claim, users must command clarity from the vendor stating how different types of defects are to be handled before entering into any agreement.

From a strictly hardware perspective, defect support is parts replacement. If a part breaks, another one takes its place. This can be done easily

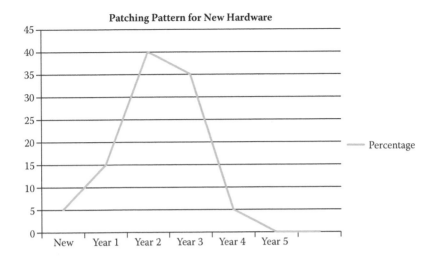

FIGURE 15.2
Patching pattern for new hardware.

in or out of warranty, provided the parts are available. In the context of FUD, the primary concern is access to the patches and fixes to the embedded software (machine code).

Figure 15.2 shows the typical machine code patching pattern for new equipment. Errors not caught in testing are only identified after new models are introduced to the field. The first few early adopters of the equipment are most likely to experience errors. However, it takes more than a few users to identify all the bugs, which is demonstrated by the peak of the patching pattern coming at the point in time when the product has been exposed to a wide range of environments. Shaking out the majority of bugs is dependent on the rate at which the model is exposed to variety. The patching pattern is not a traditional Bell curve, because patching drops rapidly to near zero as patches resolve documented problems. At this point the product becomes stable, or the manufacturer ceases patching, both of which occur long before the equipment is no longer useful.

For the majority of IT users, the period of major defect identification and patching happens within the warranty period. By the time the product version ceases production, often within 2 years, there is no further active patching because most defects will have been long ago identified and patched. The OEM may continue to offer defect support, but the types of patches provided beyond the first 2 years should be scrutinized for applicability and usefulness.

The vast majority (over 90%) of patches to machine code delivered to stable releases are to add feature support and not defect support. Buyers can require in their contracts that they have unlimited access to safety and security patches and fixes and protect themselves from unreasonable demands to continue to purchase postwarranty hardware support agreements. If there are features that can be purchased, the machine code updates will need to be provided as part of the feature purchase.

Confusing Hardware Defect Support with Software Defect Support

Most, if not all, long-term active patching is done at the operating system or application system level, which rides above, and independent of, the hardware layer. Manufacturers that provide operating systems software, including those with a custom operating system (OS), have been the most likely to conflate patching of hardware machine code with the entirely separate function of patching of their OS. Buyers that require separation of their contracts will have more favorable long-term control of their purchases than those that acquiesce to these demands.

The inverse is also bad. There are manufacturers, including some famous names, which have pulled operating system administrative functions out of their software license maintenance agreements and set them within hardware break–fix contracts. These OEMs then sell maintenance agreements as the only way to execute remote diagnostics, control of machine settings, and control of access settings for users. These are extreme lockouts for buyers that are no longer able to configure their own systems or manage their own assets without buying passwords through a noncompetitive service agreement. Some of these same OEMs have taken the further step of refusing to sell legitimate upgrades for software products without having a current hardware maintenance agreement.

Fear of Inadequate Skills

Delivery of hardware break–fix in the field requires the physical presence of a person. Many locations are not practical for a manufacturer to staff or even stock parts. Nearly every manufacturer, including the largest, uses a network of subcontractors to service some locations and some products, without being direct employees of the manufacturer. These independent service companies are the very same organizations that pitch their own service options using the same people and qualifications. The major true advantage that the manufacturer has over the independent is its ability to provide immediate access to parts and associated machine code.

In some cases, the manufacturer provides specific training and certification only to selected partners. "Certification" is often just a ruse to control the marketing of OEM service, as the skills required to perform field swap-and-replace functions are rarely unique. Most electronics today are assemblies of off-the-shelf parts. Anyone capable of dealing with one set of assembled parts, such as an Intel chip, is fully capable of handling another brand of part, such as an AMD chip. Organizations such as CompTIA have been providing a common core of certifications for millions of technicians. Much vendor-specific certification consists of reseller representation agreements and not any actual training. Many technicians and sales staff are certified by passing exams, the difficulty of which is highly variable.

Products themselves are designed to reduce the skill and training needed for the labor force. Many repairs are now executed with a simple push-pull of parts, leading to low educational and salary requirements for technicians. Networked technologies with remote diagnostics have largely replaced the diagnostic role of the technician. If there were truly advanced complexity in the repair process, OEMs would be in a hiring battle for qualified staff and would be unable to subcontract for labor, a process that is increasing and not decreasing in popularity.

Components are sufficiently similar that the skill needed to pull and replace an item such as a memory card is immediately transferrable to all memory cards. Many machines of the same general type (such as personal computers or cell phones) are assembled on the same factory assembly lines as competitive products using the same factory staff and similar, if not identical, components. There is little in the way of skill that is not as easily transferrable between one OEM technician and another.

Fear of Counterfeit Parts

Counterfeit parts are a problem throughout the entire supply chain. Measures to control and avoid distribution of counterfeit parts are being taken by all types of legitimate businesses. OEMs are not immune to stocking of counterfeit parts, which often enter the supply chain during product assembly. (For more on the topic, see Chapter 9.) Since no one can assure total purity, claims by OEMs that hint competitive sources are not legitimate are not realistic. The best policy for users is to insist that all parts, including those from the OEM, be inspected prior to shipment and suspicious parts be immediately subjected to validation or promptly replaced.

Buyers may wonder if the requirements to buy any of these contorted support agreements are legal. The reality is that it does not matter unless the customer is willing to litigate. Reluctance to go public on complaints about vendors, likely out of fear of vendor reprisal, is the number one reason that the terms and conditions being discussed are as unfavorable to owners as they appear. Other than the willingness to publicly denounce abusive vendor policies, the best any organization can do is to carefully deconstruct any myths and FUD that are presented, and then negotiate for appropriate terms.

SUMMARY

Buyers that can anticipate the marketing FUD that will be presented by the OEM will be well positioned to avoid being led into predatory or undesirable terms and conditions. There is an entire industry devoted to training salespeople to sell ice cubes to Eskimos. Technology buyers are not immune to these tactics, and those who are alert to such messaging can drive better agreements for their organizations.

NOTE

1. For the entire ruling, see "An Author of Software Cannot Oppose the Resale of His 'Used' Licences Allowing the Use of His Programs Downloaded from the Internet," Court of Justice of the European Union press release, July, 3, 2012, http://curia.europa.eu/jcms/upload/docs/application/pdf/2012-07/cp120094en.pdf.

16

Business Intelligence for Support and Maintenance

INTRODUCTION

Developing an effective support and maintenance strategy requires some knowledge of the frequency and type of problems that will occur. This is essential information for any organization for several critical reasons:

- System Availability (Risk)
- Projecting Total Cost of Ownership (TCO)
- Management (Planning and Monitoring)

The adage "One cannot manage what one cannot measure," attributed to Lord Kelvin, is completely appropriate to this chapter. There is no potential to rationalize contracts or costs without first having a set of measurements.

USES OF ANALYTICS

System Availability

Everyone understands the value of system availability to modern business. Downtime is carefully tracked and those with very little system downtime may brag about it. Business analytics specific to downtime can be used to evaluate and investigate root causes of downtime in order to reduce risk in the future. The most obvious and useful metric for measuring downtime is to track existing incidents of failure, even if they do not result in downtime. The reason that every failure is important is that in redundant environments, components fail and would have caused downtime but for the presence of a hot spare. This does not eliminate risk and may create a false sense of security.

Total Cost of Ownership

Correctly projecting total cost of ownership (TCO) includes factoring in all costs associated with keeping equipment in service for the period of useful life. Warranties rarely reflect the true total cost of ownership as internal costs to the organization are often ignored. For example, each call to the customer-owned help desk or service desk has a cost, each escalation has a cost, and the loss of productivity for the impacted end user has a cost. "Free" warranties are only part of the repair and support plan for maximizing useful life.

Following a service event from start to finish is a useful exercise to determine the additional costs to the user for warranty events. Each step in the process has a cost, which must be included in the TCO analysis. Figure 16.1 shows how TCO calculations often stop well short of the full spectrum of costs related to any service event.

Many users routinely consider the costs of the postwarranty service contracts for equipment and license maintenance contracts in TCO calculations. Most do not consider that the in-warranty period is material as

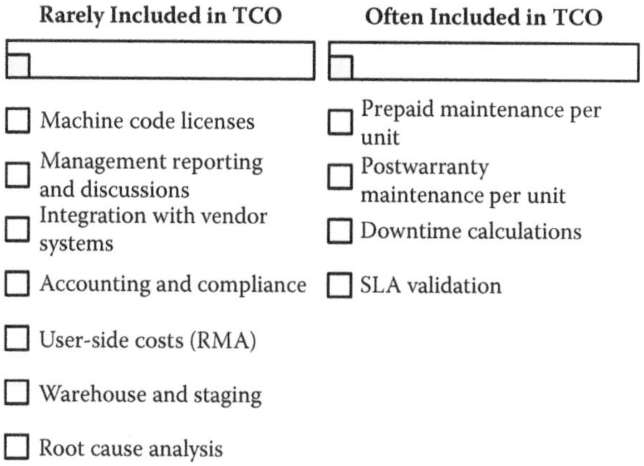

Rarely Included in TCO	Often Included in TCO
☐ Machine code licenses	☐ Prepaid maintenance per unit
☐ Management reporting and discussions	☐ Postwarranty maintenance per unit
☐ Integration with vendor systems	☐ Downtime calculations
☐ Accounting and compliance	☐ SLA validation
☐ User-side costs (RMA)	
☐ Warehouse and staging	
☐ Root cause analysis	

FIGURE 16.1
TCO for service events.

well as where many hidden costs lurk, particularly in environments where the warranty service does not include all parts and labor. Calculations are often made, but not comprehensively, of the impact of unscheduled downtime, often more as an exercise in cost justification for the original equipment manufacturer (OEM) service agreement than a real appraisal of downtime costs.

Users are also carefully focused on analytics that allow them to monitor compliance with Service Level Agreements (SLAs). This is only the tip of the iceberg in analytics since the real operational goal is to avoid failure of hardware or software. Measuring SLA compliance does not change the failure rate of the hardware or the bugginess of the software.

The big picture for TCO must include accounting for the internal hidden costs to the organization as well as the direct costs. Drilling down to associate costs for each service event is not easy and often broad guesses must be used to get the analysis rolling. If guesses are made, the analysis is still valid so long as the same guesses are made consistently for all products at the time of the analysis. Changes in the criteria must be documented and can be rerun against the earlier analysis to compare results.

There are large costs associated with systems and processes supporting repair and software service events. These must be included in the TCO because if there were no need for service, the work of the organization would be far less costly. Systems include not only the help desk or service desk, but also the work done to integrate various forms of electronic transfer of information to service contractors, procurement systems, warehouse systems, staging and test bench systems, and so forth. Even accounting and finance staff are involved with lifecycle planning and replacement cycles.

Management interactions are particularly hard to quantify because the manager is not usually compensated per event or per hour, whereas a technician or help desk employee may have more directly accessible compensation metrics.

Planning for equipment lifecycles is also informed by break–fix needs. Products that are more frequently down for repair than others should be higher on the list for replacement than more stable models. Monitoring of break–fix activity is an early warning system of aging or flawed components. Once the normal operation of the product is measured, then out of norm events can be evaluated for cause and corrective action taken.

BASIC MEASUREMENTS

The most common measurement of need for break–fix is the "mean time *between* failure (MTBF)" of any part or assembled device. One of the largest advantages of using MTBF instead of other more statistical techniques is that MTBF is a calculation using a known quantity of assets, and the quantity can be very small and still produce a result. Users can quickly find out for themselves which products in their enterprise are breaking down without having to consider if the size of their portfolio is statistically significant. More data adds *heft* to the results, but the results are still a measurement.

For purposes of maintenance and support contracting, MTBF provides actionable information with low levels of difficulty and minimal training. Most manufacturers use MTBF in their specifications for parts and testing allowing for an easy common language for discussion.

It is important to understand that MTBF is not the length of time before failures occur but is the average time between documented failures. In order for MTBF to be calculated, there has to be a reported failure, or there is no failure rate. If particular models of equipment are deployed in small quantities, the MTBF of the device may remain unknown for a long period of time. The first reported failure will complete the simple math and a rate will be calculated.

MTBF is simple math dividing the total quantity of the same model unit in use by the total failures reported during the selected time period. Most users contract for repair in months (36 months) so calculations of MTBF in months are easy to use.

As shown in Figure 16.2, the user had 100 of a particular model of servers in use for a whole year. The user had 17 repairs made, as confirmed

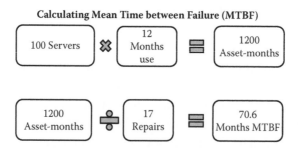

Calculating Mean Time between Failure (MTBF)

FIGURE 16.2
Calculating mean time between failure (MTBF).

by service records. (Calls for service that did not result in a repair are not included.) The resulting calculation shows that the model is running at a failure rate of one failure on average for every 70.6 months. In this example, it does not matter how many servers the vendor has sold, only the quantity of units under the reporting control of the user for a known period of time. The failure rate calculated is a reflection of actual field failure in the user environment, which is different from bench testing or accelerated testing done by the OEM. MTBF of a discrete population of devices is different as well from any larger statistical estimations of expected failure rates using sampling methods.

In an ideal world, buyers would know in advance the failure rate (MTBF) of proposed equipment in a standardized form and use that information to make simple calculations of the frequency and type of failures to be expected over the projected useful life of the equipment. As shown in Figure 16.3, if such information were available, buyers could consider products on the basis of failure rate and not just marketing hype. The four server models in Figure 16.3 show a wide range of failure rates. All things being equal, Model D, with a failure rate of 120 months MTBF is the clearly more stable product than Model C, with nearly five times more failures.

Calculations of MTBF are immediately useful to managers not only to allow for the comparison of products but also for making reasonable calculations of the need for service for that product. For example, using the failure rates shown in Figure 16.3, the user of the equipment could quickly calculate that for the highest reliability product (120 months MTBF) an

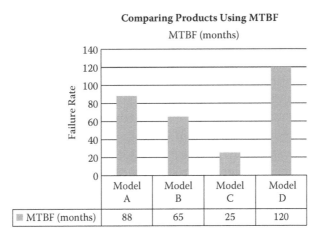

Comparing Products Using MTBF

MTBF (months)

	Model A	Model B	Model C	Model D
■ MTBF (months)	88	65	25	120

FIGURE 16.3
Comparing products based on failure rate.

installed base of 100 units would require 10 repairs per year (or roughly 1 per month). Decisions regarding how much to spend for service are much more easily made. If the OEM service agreement were to be presented for $10,000 per year, the cost per repair would be roughly $1,000 per repair. The user could then judge if that fee were appropriate for the service considering the cost of parts and labor for the device. Perhaps the service offering is $100,000 for the year, in which case the economic decision at $10,000 per repair might be untenable given the options of self-support, hot spares, independent repair, or outsourcing. The point is that drilling down to the cost per repair based on a known failure profile is enormously helpful in rationalizing service costs.

Failure rate calculations are within the reach of any organization that has a reasonably up-to-date list of deployed assets (assets in storage do not count) and a repair tracking system. Depending on the level of detail kept by the tracking system, it is possible to also evaluate and track the types of problems repaired. Even without the details, the base failure rate of the model in use in the enterprise is powerful information that is far better than anything provided by the OEM.

Users can begin calculating MTBF of their installed devices the moment they are deployed or adopt the calculation at any point during the lifecycle. Tracking products over time allows the owner to visualize the changes in failure rate that shows declines in reliability. Other than "infant mortality," the failure rate of electronics of new equipment is usually the best it will be since time and wear will only increase the failure rate.

Starting with the best case, managers can make basic projections and then monitor the equipment for changes as the product ages and adjust accordingly. In the example in Figure 16.3, for any of the server models shown, if that were the initial failure rate over the first few months, operations management could cull the repair records and see which parts were the most common points of failure. Both the user and the OEM could sit down and discuss the causes of failure and see if there was a way to mitigate problems with that particular part.

Calculating failure rate allows users to evaluate if their service agreements are appropriate for the product. There may be products that need frequent attention (such as printers or tape drives) and the cost per repair while frequent may be reasonable or excessive. Other products may have high service contract prices but need very little service. It would be impossible to negotiate with a vendor for a lower priced service agreement in either case without first knowing how much service is needed. Neither

OEMs nor independent service providers are enthusiastic about users being able to compare the price of the service contract to the need for service. The only parties that benefit, dramatically, from such knowledge are users.

There are other benefits of tracking failure other than beating up on vendors for lower cost service contracts. Tracking failure rate over time eventually generates a complete "Bathtub Curve" for the product. Most electronics in use are obsolete for applications long before they are obsolete for physical decline. For the vast majority of electronics, the backside of the *bathtub curve* is unknown. Users intending to keep equipment installed for long periods of time want to know when products are becoming physically obsolete. The only way to ascertain the true end of life for components is to measure failure rates and look for changes.

Watching for changes in failure rates does not only presage obsolescence, it is also an early warning system for widespread component failure. If the failure rate of a model of equipment suddenly spikes, this would be an indication that there is a serious component problem that may impact the entire portfolio in the short term. As the *canary in the coal mine*, one would want to have visibility to these types of problems as quickly as possible.

Every technology user has the ability to calculate MTBF using existing systems and simple spreadsheet functions. Those that do not cull their own service databases for these measurements are missing an opportunity to press for higher quality equipment rather than press for lower cost repair contracts.

SOURCES OF ANALYTICS

Manufacturers

Manufacturers test equipment all the time. Quality Control, Quality Assurance, Six Sigma type programs are keenly tuned into delivering products efficiently and without problems. As a result, most OEMs have excellent test data, but none is specifically reflective of equipment MTBF in advance of widespread field deployments. OEMs are always limited by their ability to validate specifications on parts, made more difficult by variations possible within component part suppliers, until products are actually deployed. If users feel that they are being used as guinea pigs for new products that feeling is not irrational.

Bench testing and accelerated testing labs are widely used to try to confirm specifications before introduction into devices. Not every part of every device is subjected to rigorous testing nor do such tests pick up all problems. Reliability engineers know that this type of validation is the best that can be done before the product hits the field. The real test is always the field.

It is challenging for manufacturers to keep practical records of failure rates for assembled products. Each new component part entering the supply chain carries uncertainty about performance. Examples of the variations flowing into the electronics market are available on the Web sites of companies such as Dell. Available configuration options vary on nearly a daily basis because available parts and suppliers change constantly. Purchasing truly identical product in large quantities requires a large purchase order and a specific production run. All companies in the consumer space have tremendous difficulty keeping identical parts available for more than a few months across multiple production runs.

It is also unknown how each OEM tracks or defines failure for their purposes. Quality control departments that test parts returned under warranty gleefully report "no fault found" when from the standpoint of the end user, the unit was broken. The objectives of the end user to track failure and understand MTBF are quite different from those of the OEM. The end user has to deal with the practical consequence of equipment failure and the costs to their organization in terms of downtime, potential business risk, and direct costs. For the end user, "no fault found" is an inadequate response.

Performance of equipment beyond the warranty period is further disconnected from the volume of equipment in use. Even if one accepts that all failures during the warranty period are returned to the OEM for replacement, it is obvious that postwarranty the data flow is closed off. For products sold through distribution or shipped and warehoused for any period of time, OEMs cannot begin to calculate the MTBF of their products since they do not know how many are in use.

Manufacturers are unwilling to share such information, if they have it, for several reasons. First, manufacturers with any form of field break–fix service want to prevent buyers from knowing just how profitable such service really is. If the client knew in advance how much or how little repair would be required, the client would be able to evaluate the vendor proposal against many other service options and dramatically weaken the profit margins for service.

Manufacturers are also reluctant to post their failure rate data for fear that their competitors might post failure rate data that might show the competitor to be a better buy. The flip side is that manufacturers with the J.D. Power award for highest reliability would like to use such information in their marketing. Vendors have a desire for being shown as being good and are tremendously fearful of being bad. At the present time, the OEM community does not appear to want to measure quality or reliability for the benefit of consumers.

Just because OEMs are not anxious to share their limited MTBF data does not mean it is the end of the discussion for end users. Buyers have tremendous power and can insist, as a condition of purchase, that the OEM provide the most current information on its testing, the specifications of the parts, and its warranty repair experience to date. Buyers have every reason to want to know in advance just what they can expect over the long haul before they contract for break–fix services.

OEMs can also be requested, through the power of the purchase order, to validate the performance of their equipment in the postdeployment period with reporting. If such reporting is not possible, which is common with equipment that is shipped and stored, the OEM can feed warranty repair reporting back to the end user who can then make the final calculations of MTBF using its own deployment or asset management records.

Hardware Maintenance Providers

The details kept by the provider of the maintenance contract, OEM or independent, are excellent resources to tap for failure data. In order to provide a repair, the contract holder will eventually know the make, model, and serial number (or asset tag number) of the asset, the extent of the repair including all parts used, and the time taken to complete the return to service. This data does not have to be reinvented to be used; it only needs a reporting format and some attention to detail, which can be required in the contract.

Although organizations can learn a great deal about themselves and the products they already deploy, this information is isolated and limited. It would be far more valuable to evaluate the developed statistics in context to be able to answer the question "What is normal failure for this device?" Pooling of failure reporting is the next logical step in evaluating equipment durability, suitability, and performance.

Very large organizations can create excellent cross-references for themselves using just their internal resources. Smaller organizations can create their own reporting pools within trade organizations, local organizations, and so forth to compare equipment failure. For example, if there are two locations using the same equipment but the failure rate at one location is different, management can explore the causes of the difference in order to improve overall service.

Anecdotal Sources

Many organizations rely upon sources that have no statistical basis. These include white papers, articles, blogs, conferences, and even think tanks. What passes for information is often thinly veiled marketing hype. Consider the "We're the Best" type of advertising or "We're the Most Reliable Car in Its Class" claims. Just what legitimate independent body did the research to support such claims?

Anecdotes need to be supported by measurements that have been generally lacking in the electronics industry. The widespread assumption is that such information is available, but hidden. The reality is that MTBF data is not being collected comprehensively anywhere for any electronics. That this has been tolerated for so long is testimony to the largely throwaway nature of consumer electronics and the effectiveness of the computer industry in diverting attention from reliability and competing on the basis of price performance instead of durability.

Self-Evaluation

Users of electronic equipment are in the best position to calculate the failure rate of their deployed devices for themselves. Provided they have knowledge of how many of each gadget they have in use and a list of repairs made, the calculations are simple and within the capability of anyone with an Excel spreadsheet.

Users must make sure that they have at least the following bits of information collected somewhere in their systems in order to calculate MTBF for themselves. Many users have suitable systems for some parts of their equipment portfolio but not others. Users with less than perfect information and less than comprehensive self-knowledge can still benefit from whatever limited measurements can be made under the philosophy that some information is always better than zero information. Pockets of useful data must meet at least the following minimum information.

Asset information—Total quantity of equipment in use by model. Stored assets should not be considered. Assets that are now virtualized, such as during a server consolidation, are no longer physical assets. The server on which the logical servers are hosted is the only asset to be tracked, because only those assets that can break and require a repair are meaningful.

Time—Mean time between failure requires considering time as part of the calculation. A period of time must be selected for the calculation or the result is meaningless. If there are 100 servers in use for the month of May, and 200 servers in June, the calculation of MTBF must be made separately for each month to account for the changes in the portfolio. Most information technology (IT) organizations endure considerable change in their overall portfolio of devices across even a year, making a calculation by month particularly appropriate. Less dynamic environments may be able to drive useful reporting on an annual basis.

Repair information—Almost all help desk/service desk systems assign dated ticket numbers. Users can cull these systems for completed hardware repairs and compare the quantity of repairs to the size of the deployed portfolio for any specific period of time. If there is good detail, the type of problem can also be evaluated. Part-level detail is not useful without being able to first calculate the failure rate of the unit as a whole. Most users are able to start with the overall machine failure rate as a primary metric, and then use their own records to search for specific details in the ticketing system if needed.

The largest block preventing more creative use of these systems for calculations of failure rates is the lack of attention to detail within the reporting systems. Asset management systems typically lack any form of systematic naming conventions that should be used when assets are loaded into the system. Repair reporting is often plagued with lack of details about the repair, which should be recorded when the ticket is closed. All too frequently, the repair record will record "fixed" instead of any specific action.

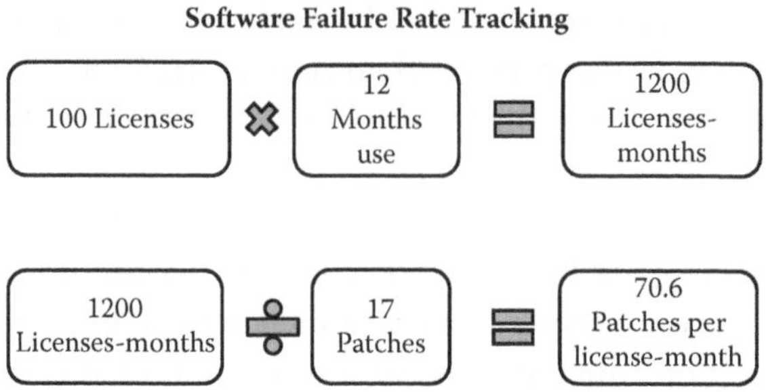

Software Failure Rate Tracking

FIGURE 16.4

Software failure rate tracking.

Software Failure Rate

It is technically possible to calculate a failure rate for any software license using the same techniques as those for the calculation of MTBF for hardware. Users could track the need for application of patches and fixes and use that as a measurement tool for contract compliance, in the same way that users track the dispatch of technicians for hardware repairs. Users could require contracts that put more burden on the vendor to provide "clean" code, or to be more discriminating about the type and purpose of patches. The key to any such efforts is to create measurements.

Figure 16.4 shows a simple calculation of a concocted measurement of "Patches per License-Month." This is just an example of many ways to compare software products based on the volume of patching. It is not difficult to make such calculations; it is only difficult to imagine how to use such information to either improve negotiations or improve the product itself. How could users benefit from calculating the relative *bugginess* of software products? Once the user has the goal in mind, the data sources for driving useful measurements can be considered and brought to bear on the goal.

USEFUL LIFE

"Useful Life" for tax and accounting purposes is not the same as physical useful life. Useful life is not empirically derived. It is very much up to the individual buyers to determine for themselves what expectations they have for the use of the equipment, and allow their tax and accounting departments to deal with useful life within that framework. Useful life in the sense of longevity in the field is determined much more by the software application than by the hardware. The majority of electronics can be kept in service in the field indefinitely if the application requires the platform remain in service. For example, there are nuclear power plants still running 1970s era hardware and software, which is clearly beyond anyone's depreciation tables at the time.

Service life is not useful life. Service life is determined by the OEM or developer, and has to do with the decision on the part of the OEM to stop supporting the product. In the case of hardware, this is the end of problem diagnosis for machine code errors. It may also mean the end of availability for previously available engineering changes or machine code updates. For software, it is the end of problem diagnosis for application or operating system code errors. It also means that the product is not going to be updated further to add support for new products, such as the native mode attachment of new disk formats or new software functions.

Deployment of technology assets in previously nontechnology products is stretching the definition of useful life in new ways. Take the case of smart meters. The "smart" part of the new meter is very much like a personal computer in a meter-shaped housing. Most of us would set the useful life of a PC at 4 to 5 years (at the outside) and business would likely depreciate the purchase over 3 years. Unfortunately, most smart meters are being deployed expecting a useful life of 20 years and a depreciation schedule of 10 years. We can all keep our fingers crossed that smart meters can be made to last in the field without extensive repair for 20 years or face big electric rate increases.

TOLERANCES AND SPECIFICATIONS

Lacking an index of product reliability over time, buyers can help identify products with a higher likelihood of success by carefully evaluating the tolerances of the equipment specifications. Electronics are highly impacted by extremes of heat and cold, of variations in humidity, and of variations in power quality. Selecting equipment that has wider tolerances should result in less failure in a wider variety of settings than a narrow band of operating conditions.

Operating tolerances are occasionally published and are often linked to warranty performance. It is common, as an example, for a cell phone to be equipped with a moisture censor to limit support claims for phones thrown into water. If this is not an acceptable limitation to support policy, the buyer can seek a different contract or a different product with a higher moisture tolerance.

Outside of the personal mobile gadget market, buyers may have to request specifications and tolerance information from their marketing team. If the sales rep does not have the information handy, it does not mean it cannot be provided. If, however, the sales rep and local sales manager does not have the information and are not inclined to push internally for the details, it should be a red flag to the negotiator that something is wrong. Vendors with high-quality products are proud to talk tech with customers about the internals of their products.

Each component part will have different tolerances and specifications. Buyers should take care to examine and understand the representations being made on high-wear components, such as disk drives and power supplies, and pay less attention to low-tech or low-impact parts. The whole machine has a failure rate that is never better than the worst performing part. So if a fan has a published failure rate of 40,000 hours, it does not matter that the clock oscillator has a failure rate of 300,000 hours. The types of parts prone to failure vary by machine but are readily observable by reviewing repair ticketing systems.

FUTURE PROOFING

There is no such thing as "Future Proofing." Many companies embarking on new applications of technology equipment would like to guarantee for themselves that they will not make an error of selecting a dead-end technology. This is best done with a crystal ball. Major new "killer apps" would not surprise us if they were anticipated a decade in advance. We would all have picked all the right stocks at the right time if we knew in advance which technologies would break through and which would evaporate.

New types of applications of technology are going to be subject to a period of shakeout of vendors and standards. The best technology does not always win. A good example of this is the Sony Betamax versus the VCR. Sony had a superior product but would not license its manufacturing to others. All other competitors then aligned with the VCR standard, possibly with some malice, and the proliferation of competition on

VCR-based products eventually dominated. The same is always likely with any new technology.

Buyers of new products in new settings should prepare to remain flexible and open to developments in the future. Short expectations of useful life are more realistic. Plans for upgrades should also include fallback plans for equipment replacements. In the early stages of deployment, some experimentation should be made with several products to gain the widest experience possible of available options.

Moore's law tells all of us that the pace of technological change is more rapid than our ability to digest and deploy dramatically improved products. Organizations that keep their options open and hold back on their commitment to a monolithic selection will be rewarded with the advantages of products not even on the drawing board but potentially transformational in just a few years.

SUMMARY

Controlling product lifecycle is best done with knowledge. Anywhere that a measurement can be used to inform decisions, the decisions can also be evaluated over time using the same measurements. Even if the full range of analysis data cannot be brought to bear on the question, partial information always trumps no information.

17

End of Service Life and Obsolescence

INTRODUCTION

Many vendors manipulate users into replacement purchases or upgrades based on the impending event of End of Service Life (EOSL). Although there are business reasons for the manufacturer to cease investing in active support, EOSL does not mean that the product (or application) itself is no longer useful or useable. EOSL most often marks the total transition of support from the original equipment manufacturer (OEM) to an independent provider.

DEFINING END OF SERVICE LIFE

End of Service Life (EOSL) is a formal announcement made by the OEM that it will no longer diagnose and correct defects in software (including embedded software and machine code) or will no longer offer hardware maintenance contracts for that model. Most manufacturers drop active support of software, including machine code, long before they drop active support of hardware. Understanding the meaning of EOSL as distinct from obsolescence is key to controlling the life of the asset (or application) in your enterprise.

The need for patching and changes to software is simply more frequent and results in major new releases more frequently than for hardware. It is also the case that software is more profitable to the vendor than hardware, so the OEM/developer typically focus sales efforts on the areas of highest profitability.

EOSL is far more meaningful for software products. It is burdensome to support multiple versions of the same product, particularly if the product design changed substantially between releases. Yet continued upgrade cycles are not welcomed by end users, as each iteration of a product often comes with new headaches and new learning curves.

In addition to driving new product sales, there is little reason for OEMs to officially drop support for older hardware. Any OEM could continue to repair parts and continue to subcontract for labor indefinitely. It is not financially burdensome to maintain the administrative functions for support of old equipment. Announcing EOSL is simply effective FUD (fear, uncertainty, and doubt) that the product will fall apart following EOSL. This is ridiculous upon consideration since product failure rate has nothing to do with marketing statements on the part of the OEM. Loss of support for hardware is not the same as for software, in that the only machine code enhancements made to older machines, if any, are made to allow attachments of new models. Once the owner has the model it intends to keep, such enhancements are irrelevant. (For more on FUD, see Chapter 15.)

Controlling the lifecycle of digital electronics always involves maintenance and support. Where OEMs dictate all the terms and conditions related to maintenance, defect support, and license transfer, there are no choices to be made. It is therefore essential for anyone wishing to control the lifecycle for any digital product to make sure that their rights as equipment owners are not subsumed by manufacturer policy.

The tendency of business-class users has been to consider a postwarranty maintenance policy only when there is a known interest in keeping equipment beyond the warranty period. Many organizations have avoided dealing with postwarranty support entirely by tying equipment refresh cycles to warranty expiration. This is both shortsighted and financially damaging. No organization can fully predict the future. If options are not negotiated during the purchasing process, they will not be available if plans change.

Moreover, tolerance of policies that are confining in the future, degrades asset value. Many buyers learn at the end of the warranty period that they have no options to resell their equipment. No investor should allow such terms and conditions on general principle. If the user truly wants to pay only for the use of equipment and not bother with asset value, the user should lease and not own.

The same policies that prevent users from selling their used assets also prevent users from deploying used equipment. One of the most powerful negotiating tools for buyers of new equipment is the availability of the same exact equipment on the used market. If there is no used market, then the OEM has much more control over the pricing (and terms) of the new equipment sale. Although it can be argued that the initial purchase choice can be a competitive platform decision, once the platform is decided, the only competition is used equipment. Users who are already entrenched with a particular OEM are entirely at the mercy of the OEM without a used equipment acquisition option.

Negotiation for flexibility of repair and support has no downside. The worst-case scenario is that the vendor declines to modify its policies. Even if the policies stand, the users will have complete knowledge of the limitations of their agreement and will not be blindsided in the future. It is very likely that with preparation and knowledge, users will be able to prevail on key points that are important to them and in so doing, make progress toward controlling their equipment lifecycles.

SWEATING ASSETS

"Sweating" is an accounting term related to stretching the use of the asset beyond its scheduled depreciation as a financial benefit to the organization. Many organizations avoid replacing products until absolutely necessary to get the most value from their investments.

Technologists routinely argue with accountants that older products are more expensive to keep in service, but often these arguments fall flat in the strictly hardware arena unless the failure rate of the older model is actually evaluated. Most often the need to replace older assets is driven by much more useful applications becoming available that will not run on older equipment.

Keeping older equipment in service is essential to sweating assets. Repair and support policies must be carefully negotiated at the time of purchase to allow extended use of technology products. Any policy agreements that allow the OEM to be exclusively involved in the support of assets postsale will be used by the OEM to force new equipment purchases. The marketplace for support and repair must be competitive or it will not be possible to sweat assets.

When repair and support policies of the OEM prevent repair, then older equipment is made obsolete by the OEM and not by the owner.

PRACTICAL VERSUS PHYSICAL OBSOLESCENCE

Physical obsolescence is largely a matter of quality control during product design and manufacturing. Engineers know how to design durable products and manufacturers know how to build them. The question is not that the materials themselves degrade, but that in the quest for lowering costs the quality of the materials specified is not a priority. If a manufacturer can shave $10 off the retail price of a tablet by using a lower quality part, and that part can hold up for just a few short years, the manufacturer is most likely to choose the cheapest practical part.

There is little or no focus on durability of electronics for the majority of business users. Corporations expect to keep equipment in service for 3 to 5 years at the outside and will not pay a premium for durability unless it can be proven that the money spent is well spent in the short term. This is much more realistic than most understand, but the combination of fast refresh cycles, reliance on redundancy instead of quality, and general shortsightedness all contribute to drive the industry toward price and not quality.

There is little evidence that well-built electronics physically age into obsolescence in less than a decade and likely a good deal longer. Reporting culled from TekTrakker® indicates that many devices remain in the field for several decades provided that service parts are available. It is also observable that older models of equipment made in the United States are far more durable than their replacement products manufactured under price pressure in Asia. For example, an HP laser printer of the 1990s was very sturdy and could be placed on a factory floor and subject to significant rough handling without breaking. The replacements, while cheaper, use plastics instead of metals and lighter springs and clips, and as a result cannot survive any rough handling and need more service for problems with paper paths than the older models.

Most equipment replacement is done to support applications that do not function on older models of equipment. Larger memory sizes, faster processors, and faster networking connectivity are needed to support applications and operating systems that are written without a requirement to run on older machines. There is little premium today placed on *tight code* except for devices, such as cell phones, where power management (also known as battery life) is at a premium. Tightly written machine code programming for cell phones requires fewer processor cycles and therefore less power to operate. This in turn allows the manufacturer to market higher battery life, which is a value point for consumers.

Applications advances that drive widespread computing platform replacement are called "killer apps." Some of the earliest killer apps that prompted personal computer hardware sales were word processors and spreadsheets. Early networking applications drove many business sales of terminals (not personal computers) for data entry later followed by replacement of "Dumb" terminals by personal computers often emulating the work done on terminals but with higher speed attachments. These older personal computers were not set up for Internet access, which had to be replaced yet again to provide the necessary interfaces.

Operating systems can also be *killer apps* by offering valuable new functions such as enhanced security or embedded networking. (Internet Explorer versus Netscape browser wars were fought over Microsoft deciding to package Internet Explorer as part of its operating system.[1])

New equipment features, such as networking attachments, CDs and DVDs, and enhanced displays drive replacements. Most of us have electronic equipment in our homes that is perfectly functional but gathering dust in a corner. TVs without HD, stereos without an iPod jack, laser disk players, floppy drives, and so forth, are no longer practical to use. The same holds with business-class electronics. The advantages of new equipment features quickly obsolete older models.

Whatever new killer apps await, displaced products will be functionally obsolete long before they die of physical causes. These functional but low value assets have to be physically removed to make room for the new equipment, thus feeding the market for recycling and repurposing of used electronics.

REFRESH AND REPLACEMENT PROGRAMS

Rolling replacement programs (Figure 17.1) have been the norm for fast-moving technology such as personal computers and mobile devices. These programs often replace equipment based on the expiration of the OEM part warranty, even if the labor is provided by independents. This same mind-set is also dominating cell phones with the carriers building in the "upgrade" to new models at the end of each contract period.

Rolling Replacement Cycle

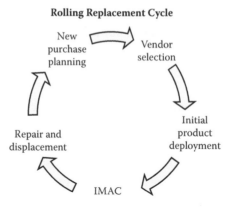

FIGURE 17.1
Rolling replacement cycle diagram.

These replacement cycles are entirely arbitrary. Most are pressed by OEMs for the obvious reason that OEMs want to sell more equipment more often. Users miss the opportunity to sweat functional assets for extended periods of time, and at the same time miss the opportunity to proactively replace problematic equipment. This process can be made more rational by tracking the failure rate of equipment and taking action based on need.

Even without a formal program, need-based replacements are practical anytime that a unit needs repair. The labor involved in swapping spares is the same as swapping the old model for a newer unit. This process takes advantage of the costs already built into the logistics and reverse logistics of product shipment and recovery, plus it leverages the administrative work already involved in tracking the shipment and deployment of a new serial number of the same model. At this point, changing the model is very cost-effective.

Failure-Based Refresh Programs

Users with no compelling need for new equipment can take advantage of the opportunity to keep their assets in use until they fall apart. No one actually wants to wait for catastrophic failure before replacing equipment, so most managers make an arbitrary determination based on age and replace equipment *just in case* it might be getting too old.

The ideal situation for users is to replace equipment just ahead of any surge in failure rate. It must be understood that the perfect moment in time for replacement cannot be known without measurement. Owners that track the failure rate of their equipment portfolio can review the failure history of equipment and monitor the reporting for changes. If, for example, a portfolio of 4-year-old blade servers started showing an increase in motherboard failures, it would be worth doing extra investigation on the root cause of the board failures. Perhaps aging capacitors are at fault, in which case the product should be considered for replacement.

Leaving the calculation of failure rates to the OEM is not only impractical but also foolish, since the OEM has a natural and appropriate vested interest in selling replacement equipment. OEMs do not want to be selling extremely durable equipment and will only improve durability if they must prove durability to make new sales. If asked, OEMs will not encourage the use of older equipment, even their own equipment, and will always take the question as an opportunity to sell replacements.

If users are going to ask the OEM how long will this stuff last, then the user should expect the response of "we don't know." If there is an empirical response, the user should be asking the OEM "how do you know that?" There are few ways for OEMs to make legitimate longevity claims about their equipment, only one of which is documentable. The OEM does know about failures reported under warranty or under postwarranty service contracts. Unless the product in question is known by the OEM to be substandard under its criteria, the point of physical obsolescence is extremely unlikely to be observable during the first few years of use.

As shown in Figure 17.2, the OEM has limitations on both the quantity of assets in use and the full spectrum of repairs. Each OEM knows in excellent detail how many units of a model were shipped, but few know how

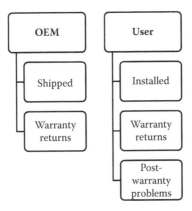

FIGURE 17.2
Warranty and postwarranty tracking.

many were actually deployed. Very few OEMs even know how many of their sold assets are in use even from the moment of shipment to distributors. Equipment on the shelf at retailers, in storage in warehouses, or in storage at an end-user site cannot be differentiated from equipment in use.

The OEM is also disconnected from failure reporting, as only returns made under warranty are visible to the OEM. If the user does not make a timely return or chooses to self-repair, the OEM will not know of all problems that occur even during the warranty period. Postwarranty maintenance is far less visible to the OEM, since the OEM does not directly repair many products. The older the equipment, the less the OEM knows as more and more devices remain in the field beyond the expiration of the initial warranty. As a result, the OEM rarely knows the failure rate of their products over time and can only estimate. Owners are the only parties that can track failure rates over time.

Warranty-Based Refresh Programs

The fundamental problem with warranty-based refresh programs is that warranty options for digital electronics are not related to the durability or quality of the equipment. Warranties, as explained in Chapter 3, are a marketing tool and as such are negotiable. OEMs use the timing of the warranty to enhance their sales, not to prevent problems for the end user. Once committed to a refresh program with a particular vendor, the selection of products for the next round of replacements is invariably limited to those of the original OEM. Refresh programs can be created for terms other than those initially offered by the vendor and can be administered by parties other than the OEM.

Such refresh programs do not require that only a few models of equipment are deployed, but this is also common practice. Managers strongly prefer to have consistent configurations of hardware and software deployed in the field. (The field is anything other than the locked data center.) It is far easier, and presumably less costly, for the help desk/service desk to support a well-tested and well-understood set of equipment. This reduces the number of unknown interactions between products, particularly the software stack, so long as users do not add their own equipment or software without the knowledge of the support team.

Corporate buyers of digital assets distributed to employees have for many years adopted formal refresh programs so that their portfolio of devices is replaced on a regular basis. Such programs frequently match the refresh cycle to the warranty coverage offered by the manufacturer. A common formula has been to replace laptops (and other mobile assets) on a 24-month cycle and deskside assets on a 36-month cycle.

These programs work well in theory but have problems in execution. Users, particularly mobile users, are not gentle and often physically damage assets. Warranties do not cover damage, unless separate insurance is purchased. It is often difficult to force employees to pay for damage to company-owned assets, so the portfolio manager usually needs to provide for repair services outside of warranty under a corporate agreement.

Rolling Refresh Programs

Rolling refresh programs and rolling deployments all suffer from the practical problems of component variability. It is difficult for sellers to continue to produce identical products over a long period of time. In the world of digital electronics 18 months is a long time. Most users have come to accept a modest level of component change through deployment and refresh programs in return for consistent pricing.

This only provides the illusion of consistency rather than the fact. When it comes to repair and maintenance, small differences are just as important as large ones.

Competitive bidding for repair services for personal computers is common, but restrictions from the OEM on access to parts, manuals, schematics, and so forth must be avoided if repair programs are to be truly competitive. (See Chapter 7 for details about the essentials needed for repair programs.) Buyers need to consider as well that policies that are common today may disappear overnight as OEMs decide how they wish to profit from their own service offerings. For example, in September 2013, HP made a policy change blocking independent repairs that impacted 100% of users of its HP Infinity and HP 9000 lines regardless of how long ago the equipment was purchased.[2] Unhappy users may complain or even litigate. The lesson learned is that policies that might never happen must be prohibited in the initial purchase agreement or users may have to press their rights in court with unknown outcome.

Bring Your Own Device

"Bring Your Own Device (BYOD)" is highly problematic for management of repair and support. Best practices for BYOD have not yet settled, nor will they necessarily settle. The pace of change in the area of mobility computing is accelerating and probably should not be expected to stabilize. The following issues are already known to exist in general terms:

- Privacy
- Security
- Repair
- Reimbursement
- Tracking

Privacy issues are at the cutting edge of digital rights, but are not repair and support rights. Owners of equipment in many areas, including telematics for automobiles and smart meters in the electric grid, are not clearly in charge of the data flowing off the device. In the same way, privacy issues abound over the use of e-mails and other insecure Internet traffic by government agencies or others.

Security, in the context of BYOD, is largely the area of providing individuals with a secure link into a corporate Web site using a VPN (Virtual Private Network) or other capability. It is also a widespread security concern that many mobile devices can be easily infected by malware, which may interfere with the ability of the employee to use their device or be used to disclose company data that might reside on the device. In the physical sense, security also means the ability to shut off traffic to an employee-owned device if the employee is terminated or if the device is lost or stolen.

Repair of employee-owned devices is not efficient. If the employee needs to seek his or her own repair contract or services (as is common with Apple products), the individual buyer is not getting the advantage of a volume agreement that would otherwise be negotiated by professionals. Although the corporate office is relieved of the task of buying and supporting individual devices, the corporate office is often reimbursing employees for the use of their devices, including the insurance and service plans that are individually bought. The net cost is not likely a savings.

Not keeping track of employee-owned devices is an advantage to the corporate buyer, as the costs to implement an asset tracking and service desk function for mobile assets may easily exceed the value of the device. This, more than anything else, is a real financial driver for permitting BYOD so long as security issues are not a pressing concern. When employer-issued laptops were $5000 each, it was difficult to justify holding the employee accountable for the device. At $500 each (or less), the tracking and management costs to care about such assets are potentially backward in terms of return on investment (ROI).

SERVICE PARTS FOR THE LONG HAUL

Many products remain in use successfully for decades because their application requirements are unchanged. There are many markets where very old equipment is still in use, such as nuclear power plants, and more modern markets, such as smart meters, where products are expected to be in the field for a decade or longer. Tactics used to support equipment for the long haul fall into three major categories: scavenged parts, board-level repair, and custom manufacturing.

Scavenging is a common source of parts not only in the long term but also in the short term. If OEMs do not provide service parts through their parts desk or if the pricing is too high, repair companies will purchase parts off used machines. OEMs scavenge parts because of availability problems. Not all OEMs carry all parts at all times in their warehouses. Users that are told they are getting only "new" parts are probably not being told the truth, particularly in the later stages of an extended warranty.

Eventually, the supply of scavenged parts is insufficient to support the repair needs of the product. The next level of repair support is provided by specialist companies that perform board-level repair. These firms support very old equipment as well as not-so-old equipment for manufacturers, independent repair companies, and directly for end users. The majority (upward of 90%) of electronics can be repaired at the board level, provided of course that the product was not completely smashed. The board-level repair option is part of the entire ecosystem of extending the useful life of all electronic equipment.

By looking closely at a circuit board one can see the individual parts and connections. Failed parts can be identified through testing and replaced, or loose connections reattached. It is also possible when working at this level to upgrade componentry, so long as the microcode for the part is available. This type of work has a high labor cost for obvious reasons. In many cases, the only way to keep assets in service is to arrange for this type of repair.

The alternative to board-level repair for older electronics is a custom build of a replacement part. This is done frequently but rarely for products that are not in use by the millions of units. It is expensive to set up a manufacturing plant for small quantities (under 1 million units is considered small in the industry).

In addition to limitations on patents, custom manufacturing may be impractical for very old parts because the manufacturing equipment needed to build the part is itself no longer in use. Every major product iteration in chip building causes a massive wave of retooling of manufacturing lines to accommodate the newer products. Losing the ability to manufacture replacements en masse makes the support of old equipment an exercise in antiquing. Just as the Ford Motor Company no longer manufactures parts for the Model T, the same problem of parts availability apply to digital items.

SECONDARY MARKETS

Extending the useful life of digital devices is tied to the ability to trade products in the secondary market. The principles that support repair also support used equipment trading and vice versa. One market does not exist without the other.

Principle: Used Equipment Is the Best Source of Parts outside the Manufacturer Service Desk

Buying and selling of used equipment is a time-tested method of acquiring service parts in all industries, not just computing. Many times the parts are more valuable than the whole, although this is more likely true in the auto industry than in personal computing. Even when the manufacturer has set limitations on the transfer of software licenses that would need to transfer to a secondary buyer of a whole machine, parts rarely have the same restriction.

The main reason that parts trade without licenses is that parts are rarely tracked as independently serialized assets. This may change with improvements in the ease of tracking using other digital technology, but the net benefit to the manufacturer to make such investments would have to be substantial. Most asset tracking as it exists today, for both manufacturer and user, is geared toward associating configurations, costs, and licenses with the frame or enclosed as the major asset. The internal parts are often pulled and replaced during the service process, so tracking changes in seri-alized parts is labor-intensive and may not deliver any value for the effort.

Whole machines are often subject to restrictions on software license transfer, including limitations on transfer of embedded software, oper-ating systems, and application licenses. Depending on the license policy, whole machines may be impossible to trade in the used market. If whole machines cannot be bought and deployed, the chances that older mod-els of equipment will remain in use are also diminished. The few older models in use, the fewer parts are needed for postwarranty or non-OEM support, and the parts values decline in value in tandem with the limita-tions on the trade in whole machines.

Principle: Broken Machines Do Not Sell Unless Repairable

Old information technology (IT) equipment can be broken and still have value, so long as it can be repaired. The seller of a broken machine can pay to have the machine brought back into service or the buyer can do the same, and still have value. The challenge comes if the OEM refuses to repair equipment or will not allow others to repair equipment. In these cases, any broken machine has zero value and any machine that might ever break can be similarly worthless. As soon as the OEM commands all repair, the equipment only has value if the OEM decides to let it have value.

Terms and conditions that damage used equipment value are the same as those that restrict repair, for slightly different reasons.

- The OEM will not repair equipment on a time and materials (T&M) basis. The T&M repair option is important to the used market trans-action as a way to bring older products up to date with embedded software or to make repairs that are needed at the time of installa-tion. Many machines in the used market sit in storage and may also suffer from transit damage. If the OEM will not "recertify" or repair units (including bringing machines up to current engineering change [EC] levels), the pool of prospective buyers diminishes and values decline.[3]
- The OEM will not repair equipment outside of an OEM service con-tract. The OEM can selectively price service contracts to dissuade users from deploying used equipment. Prospective buyers will take the cost of the OEM agreement into account when purchasing and deduct it from the value of the machine. Unless there is still a substantial price

advantage, the used sale will not happen. The price of a service contract, as opposed to a single T&M repair, is intentionally substantial.

- The OEM will not accept used equipment under any service contract. The OEM can completely block the installation of used equipment and thereby crush the value of all used equipment.

- The OEM will not sell service parts and associated documentation to the owner. The OEM can attempt to block users from performing self-service or engaging with independent service providers by refusing to sell service parts. This is not the most effective limit on used equipment trading as scavenged parts may be available, but it adds another level of difficulty for the user who is seeking to keep older machines in use.

- The OEM blocks access to restoration or updates of embedded software (machine code) including diagnostic routines. Prospective buyers of used equipment are unlikely to purchase equipment that is not up to current EC levels (meaning current levels of machine code). Such blocks also impact the ability of the equipment owner to buy used parts and add additional equipment to extend the useful life of already purchased equipment.

RECYCLING AND REPURPOSING

Moore's law has a major impact on the secondary market for predominantly digital assets. The rate of change between generations of equipment is so large that most IT equipment is technologically obsolete in less than 3 years. As a result, most used digital equipment drops in value dramatically, often to the point where a 3-year-old item carries less than 10% of its original value. Most asset value of used equipment is supported by the difficulty and expense of replacement rather than the actual residual value of the unit. Lessors rely upon "In Place Value" as part of their calculations of making a return on leasing equipment with very low actual used value.[4]

For those assets that Moore's law has left behind, most organizations engage in some form of "Asset Recovery" process. It does not matter if the process is formal or informal. The goal is to physically get rid of the older equipment to make room for new models and to take the equipment off the books with as much value recovery as possible.

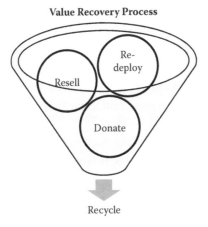

Value Recovery Process

Recycle

FIGURE 17.3
Value recovery process.

As illustrated in Figure 17.3, everyone tries to sell assets as whole machines to a secondary market buyer. If that does not pan out (and it rarely does), donations or employee sales are considered. Recycling is the last option but is also the most common due to the difficulty of extracting asset value from the resale or donation options.

Used Equipment Sales

Many products have resale value in the secondary market. The difficulty for users is finding a buyer for a package of assets rather than a single asset at a time. Most organizations do not have an interest in selling surplus one item at a time, particularly when the unit value of assets may be in the single dollars. Although Internet auction sites have made the process vastly more efficient, the owner still has to register on the auction site; post listings; respond to questions; set up payment processing; and then clean, prep, package, and ship products to individuals.

There are other issues as well that cause corporate owners to avoid dealing with small buyers, including employees. Many people prefer to sell used cars to total strangers than to relatives because they do not want to have any implied responsibility for the car postsale. This is a frequent problem with employee sales of surplus products as well. The internal help desk is still called when the employee has a problem with a surplus PC used at home. It is difficult to sever personal relationships even if the sales document states otherwise.

Users then find themselves trying to sell assets into a marketplace that is probably flush with the same assets coming into the market at the same time. If a killer app is causing the marketplace to adopt some new gadget, the chances are excellent that there will be a market glut of the products being replaced. Users are also in competition with leasing companies with the same products being returned off lease. It is often difficult to find buyers under these conditions.

Regardless of the selection of the used buyer, all sellers need to arrange for removal of user data, removal of nontransferable licenses, and deletion of any data that may be in different forms of memory other than hard drives. It does not matter who does the work, so long as it gets done.

Donations and Employee Sales

Once sellers have given up on the used market, the next options usually considered are donations to charities or employee sales. Both of these options require the owner to scrub data off equipment, remove licensed products that do not transfer with the machine, and clean and package products for shipment or pickup. If local charities are involved, many donors may be embarrassed to give away equipment in poor condition. Unless the donor is really prepared to be a support service to the charity, most corporations do not actually use this model directly.

There are services available to manage donation and employee sales programs for a fee. The seller benefits by keeping equipment out of the waste stream but does not generally get any meaningful financial recovery. Some of the vendors that provide these services also offer recycling programs and some do the work as subcontractors to the vendor providing the replacement equipment. The suite of services typically offered includes all data wipes, restoration of the original operating system, physical cleaning, minor repair, and audit trails.

Recycling

Ideally, electronics are recycled and not landfilled. Depending on the product and the state, laws may prevent throwing electronics in the trash. Many jurisdictions mandate or at least offer trade-in programs from manufacturers to keep IT equipment out of landfills. These efforts only scratch the surface of handling the disposal of electronic waste, also known as "e-waste."

There are several nuances regarding recycling that merit discussion from a services and support perspective. First, few reasonably modern electronics have enough precious metals to cover the costs of processing and handling for recycling. As a result, most recycling programs involve processing fees and not the other way around. Owners of large volumes of electronic equipment should consider disposal as part of replacement contracts, if only to save processing fees as the labor to bring a new unit in and put the old unit in a box is only marginally different. Vendors frequently offer such replacement services, which are mostly subcontracted to specialists with the logistics skills and transportation relationships.

Second, there is not a lot of business risk in disposal of machines in landfills unless they include hazardous materials. Most modern electronics are not classified as hazmat and therefore can be transported conventionally and landfilled if local regulations permit. Although the vast majority of conventional CRT terminals are no longer in use (functionally obsolete), the glass inside the CRT is coated with significant quantities of lead and cadmium and is hazmat. Specialists who can separate the glass from the other materials and properly dispose of the glass must process CRTs. Most modern electronics have been manufactured specifically to avoid the use of hazardous materials.

The larger risk to business is that inadequately processed storage media is retrieved and sensitive data is recovered. The degree of sensitivity usually dictates the processing method selected for the destruction of memory or storage media. There are a variety of methods in common use and there undoubtedly will be others developed to streamline the process or manage the unique data destruction needs of media that are yet to be invented. Currently, the average business user is most interested in keeping ordinary business data from getting into the hands of competitors. This information is rarely momentous or interesting and can be kept from prying eyes by using very simple reformatting of media. Multiple passes of reformatting make reconstructing data increasingly difficult. A three-pass system of overwrites is considered a "DOD Wipe." DOD stands for Department of Defense and its standard for low-level security data wipes is usually considered sufficient for most commercial purposes.[5]

Wiping, however executed, is the only method that preserves media for reuse. Users can save expenses considerably by wiping disk drives if the product has no market value and will not be reused or repurposed. If the product can be repurposed, it is often because the original disk media is functionally obsolete and can be destroyed and new replacement media can be installed.

The destruction of original media and new product replacement is potentially cost-effective as methodically wiping disk drives is labor-intensive. Each machine has to be running and a technician has to repeatedly run software on each machine to complete each wipe. Multiple wipes add technician time. The potential buyer of a used machine will also value the improvements in storage capacity provided by newer versions of storage media, possibly more so than other improvements that would be less obvious.

Third, the market for recycled electronics is complex. Plastics and metals from frames and cases have a value as raw materials. Tiny screws have their own value. Component parts may have value as either service parts or as parts for repurposed equipment. Although the business or consumer in the United States or the European Union would not consider a purchase based on reclaimed materials, much of the world is delighted to have low-cost options. Unfortunately, the market in used components is a source of raw materials for counterfeiting as well as for legitimate uses. Detecting counterfeits from the myriad of worldwide *chop shops* is a serious problem for defense electronics in particular and is expected to be an emerging problem for corporate buyers.[6]

Most of the costs to recycle equipment in the United States and the European Union are borne by the owner. Owners pay to have their equipment picked up and transported to a facility that will disassemble the unit, remove and carefully process any media according to specifications, and account for the disposal of component parts sold intact. In many cases, the costs of transportation to the recycling facility substantially exceed the value of the equipment as scrap even if no other services are needed.

Circling back to the discussion of the impact of recycling for commercial contracting, the message is that the disposal of equipment is an important cost element in evaluating total cost of ownership. Equipment that can be resold as whole machines, perhaps with replacement media, is more valuable than any equipment that is sold for its scrap value as metal and plastic.

SUMMARY

Buyers that anticipate how they might support equipment and applications beyond EOSL can negotiate far more effectively than those that ignore the opportunity to substantially extend the useful life of their assets.

NOTES

1. See "Browser Wars," *Wikipedia*, http://en.wikipedia.org/wiki/Browser_wars, for more information on the series of skirmishes over networking.
2. The announcement is available at: http://customer.hp.com/w/webView?cid=211097 68820&mid=1238278552&pid=2120834&vid=13597&ee=am9lc2lwcGxlQGdtYWls LmNvbQ__&si=&mv=H&bv=H&oc=H&sc=&k=19pg6O.
3. For many years, IBM had a policy of "Banding" used equipment at the request of the owner at deinstallation. So long as no one broke the band, IBM would return the equipment to service under its contract at another location without argument or additional cost. Banded equipment could stay in a warehouse for an indefinite period of time and still remain maintenance eligible. Banded equipment was the standard for used equipment trading around the world.
4. In Place Value is a combination of the expected costs for the lessee to meet the return requirements of the lease including costs of deinstallation (typically required to be by the manufacturer or guaranteed eligible for service by the manufacturer), packing, crating, and shipping to a warehouse location specified by the lessor. This is in addition to any charges that would apply based on condition problems with the equipment. In these situations, the lessor is seeking maximum reimbursement for these charges.
5. *Wikipedia* provides a reasonable discussion of wiping options at: http://en.wikipedia. org/wiki/Data_erasure.
6. See John Villasenor and Mohammad Tehranipoor, "The Hidden Dangers of Chop-Shop Electronics," *IEEE Spectrum Magazine*, http://spectrum.ieee.org/semiconductors/processors/the-hidden-dangers-of-chopshop-electronics, for a current review of the situation.

Index

F